ELECTROANALYTICAL METHODS

ELECTROANALYTICAL METHODS

S Rani

Lecturer
Department of Chemistry
Sri Chandrasekharendra Saraswathi Viswa
Mahavidyalaya University
Enathur, Tamil Nadu

MJP PUBLISHERS
Chennai 600 005

ISBN 10: 81-8094-045-4
ISBN 13: 978-81-8094-045-3

Cataloguing-in-Publication Data

Rani, S (1969 –   ).
Electroanalytical Methods / by
S Rani. – Chennai : MJP Publishers, 2008
xii, 174 p. ; 21 cm.
Includes Glossary, References and Index.
ISBN 81-8094-045-4 (pbk.)
1. Chemistry, analytical. I. Title.
543 RAN MJP 044

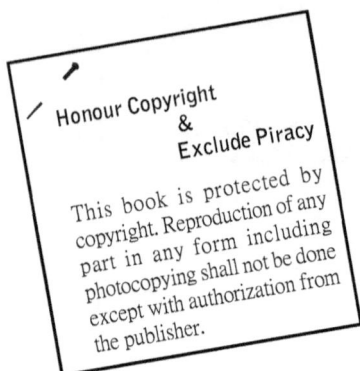

ISBN 81-8094-045-4   **MJP PUBLISHERS**
© Publishers, 2008   47, Nallathambi Street
All rights reserved   Triplicane
Printed and bound in India Chennai 600 005

Publisher : J.C. Pillai
Managing Editor : C. Sajeesh Kumar
Project Editor : P. Parvath Radha
Acquisitions Editors : C. Janarthanan
Assistant Editors : B. Ramalakshmi, S. Revathi
Composition : Lissy John, N. Yamuna Devi
Cover Designer : Lissy John
CIP Data : Prof. K. Hariharan

This book has been published in good faith that the work of the author is original. All efforts have been taken to make the material error-free. However, the author and publisher disclaim responsibility for any inadvertent errors.

PREFACE

The concepts of electrochemistry and modern methods of electroanalytical techniques find applications in all fields of science viz., earth science, life science, physical science, biochemistry, pharmaceutical chemistry and so on. Therefore, a sound knowledge of electrochemistry has become a necessity for any chemistry graduate.

The concepts have been chosen in such a manner that the book will be useful for the undergraduate and postgraduate students of Indian universities. The concepts have been illustrated with examples and problems for easy understanding. The references provided would be useful for further exploration of topics dealt in the text.

This book would enable the readers not only to understand the concepts of electroanalytical methods but also to update the advances of electroanalytical methods and use this effectively to meet the challenges in research.

I wish to take this opportunity to sincerely thank my friend Miss. M. Aulice Scibioh who inspired me to write this book. At this juncture, I also wish to thank all my colleagues and teachers.

I thank MJP Publishers for their interest in publishing this work and for the support they have rendered in transforming my manuscript into a complete book.

Comments from readers for improving the book are cordially invited.

S Rani

CONTENTS

INTRODUCTION

Electroanalytical chemistry involves the analysis of chemical species (analyte) through the use of electrochemical methods. Generally, we monitor alterations in the concentration of a chemical species by measuring changes in current in response to an applied voltage with respect to time using electrodes. According to Faraday's law, the charge is directly proportional to the amount of species undergoing a loss (oxidation) or gain (reduction) of electrons.

$$Q = n F e$$

where,

Q is the total charge generated (coulombs)

n is the number of moles of a species undergoing oxidation or reduction

F is Faraday's constant (96,487 C/mol)

e is the number of electrons per molecule lost or gained

Current is the change in charge as a function of time.

$$I = dQ/dt$$

Thus, the current response with respect to time (voltammogram) gives information about changes in the concentration of the species of interest. Electroanalytical methods such as conductometry, potentiometry, cyclic voltammetry, stripping voltammetry, differential pulse polarography, and chronoamperometry are not only capable of assaying trace concentrations of an electroactive analyte, but supply useful information concerning its physical and chemical properties such as oxidation potentials, diffusion coefficients, electron transfer rates, and electron transfer numbers. Moreover, electroanalytical methods can be combined with spectroscopic techniques *in situ* to provide information concerning molecular structures and reaction mechanisms of transient electroactive species. These methods find application in the field of chemistry, biochemistry, biology, environmental studies, physics, and medical science and also in many branches of science.

1

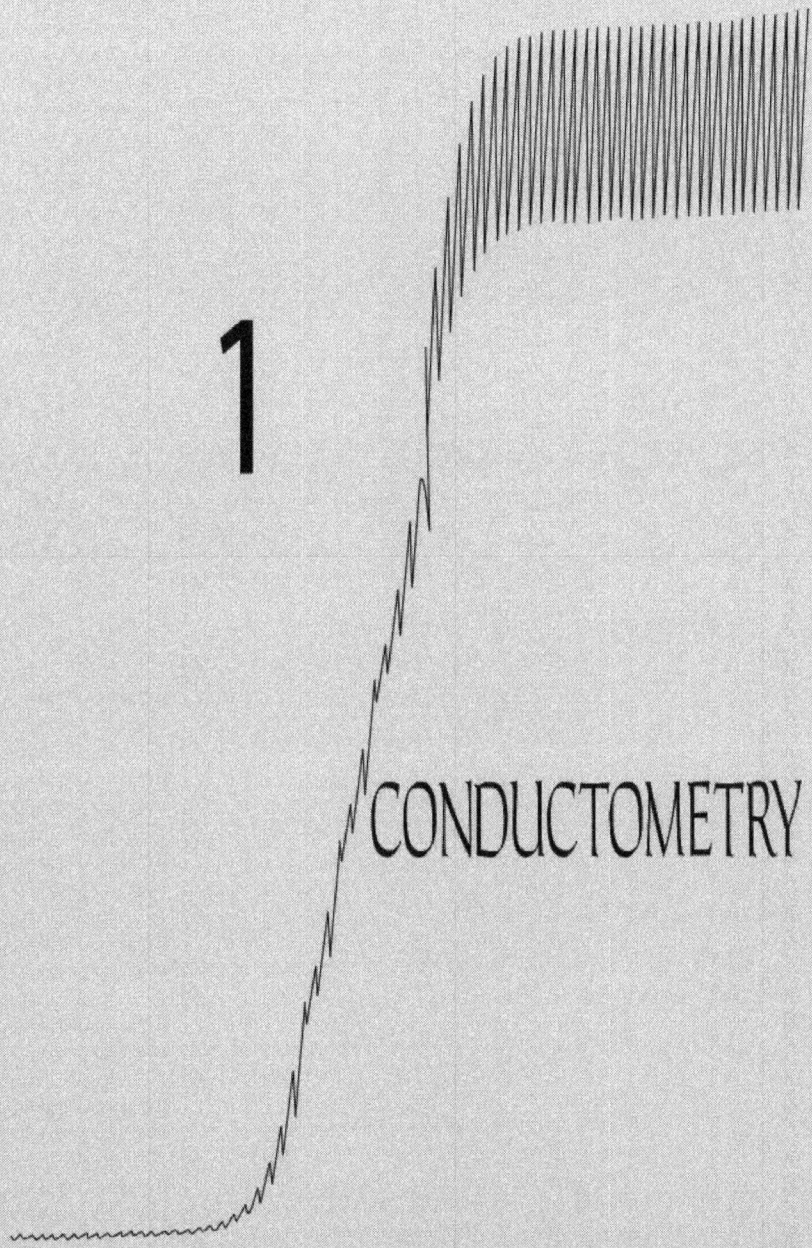

CONDUCTOMETRY

ELECTROLYTIC CONDUCTANCE

The conductance of a solution is a physical property that can give important quantitative information regarding the composition of the solution. It is a measure of the ability of a solution to carry current (migration of ions) and depends on the concentration, mobility and charge of ions in solution, and on temperature. The movement of the ions is related to the temperature of the solution and the viscosity of the solvent. As temperature increases, ions move faster and conduct more current. An increase in the viscosity slows down the movement of the ions. The dielectric constant of the solvent affects the electrolytic conductivity. The measurements of electrolytic conductance provide quantitative information regarding the concentration of specific species. In addition it also provides information about the composition of water.

Ohm's Law

Ohm's law states that the strength of current flowing through a conductor is directly proportional to the potential difference (E) applied across the conductor and is inversely proportional to the resistance (R) of the conductor. This is given by

$$I = E/R$$

where,

I = current strength expressed in ampere,

E = potential difference expressed in volts and

R = resistance difference expressed in ohms.

This law obeys both metallic and electrolytic conductions.

Specific Conductance

The resistance (R) of a conductor varies directly as its length (l) and inversely as its cross-sectional area (a). The reciprocal of resistance $\left(\dfrac{1}{R}\right)$ is called as conductance and its unit is ohm^{-1}, mho or siemen, S.

$$R \propto \frac{l}{a}$$

$$R = \rho \frac{l}{a} \qquad (1)$$

where, ρ is a constant depending upon the nature of the material and is called the specific resistance of the material.

If $l = 1$ cm and $a = 1$ cm^2, then $\rho = R$. It is the resistance of 1 cm^3 of the material. The reciprocal of specific resistance is called specific conductance (κ). It is defined as the conductance of 1 cm^3 of a material.

$$\rho = \frac{a}{l} \cdot R \qquad (2)$$

$$\kappa = \frac{1}{\rho} = \left(\frac{l}{a}\right) \times \left(\frac{1}{R}\right)$$

$$= \left(\frac{l}{a}\right) \times \text{conductance}$$

The unit of specific conductance is ohm^{-1} cm^{-1}.

Equivalent Conductance

The conductance of an electrolyte depends upon the temperature and the concentration of the solution as well.

In order to obtain comparable results for different electrolytes, it is necessary to take solutions of equivalent concentrations, i.e., solutions which are capable of furnishing ions carrying the same total charge of electricity. The conductivities of different electrolytes can be compared by measuring the equivalent conductance.

Equivalent conductance λ_{eq} is defined as the conductance of all the ions produced by one gram equivalent of an electrolyte in a given solution.

In order to derive a relationship between specific conductance and equivalent conductance, let v cm^3 of the solution containing 1 gram equivalent of an electrolyte be placed between two large electrodes spaced 1 cm apart.

Let the measured conductance of the electrolyte be C. As the solution contains 1 gram equivalent of electrolyte, the measured conductance is equal to the equivalent conductance. Hence,

$$\lambda_{eq} = C$$

The specific conductance can be given as

$$\kappa = C/v$$

$$= \Lambda_{eq}/v$$

$$\kappa_v = \lambda_{eq} \tag{3}$$

where, v is the volume containing 1 gram equivalent of an electrolyte.

If N is the concentration of solution in gram equivalent per litre, i.e., normality of a solution, then the volume of solution containing 1 gram equivalent of electrolyte can be calculated as

$$v = \frac{1000}{N} \ cm^3$$

Substituting v in equation 1.3, we get

$$\lambda_{eq} = \frac{1000\,\kappa}{N}$$

The unit of equivalent conductance is ohm^{-1} cm^2 equivalent^{-1} or S cm^2 equivalent^{-1}.

Molar Conductance

It is defined as the conductance of the solution containing gram-mole of the electrolyte. It is denoted as λ_M. It can be related to specific conductance as

$$\lambda_M = \frac{1000\,\kappa}{M}$$

where, M is the molality of the solution. The unit of molar conductivity is ohm^{-1}cm^2 mol^{-1} or S cm^2 mol^{-1}.

DETERMINATION OF ELECTROLYTIC CONDUCTANCE

We know that the conductance of the solution is the reciprocal of resistance. Therefore, if the resistance of the solution is measured, the conductance can easily be calculated. The Wheatstone bridge method is generally used for the measurement of resistance. A schematic representation of the apparatus is shown in Figure 1.1.

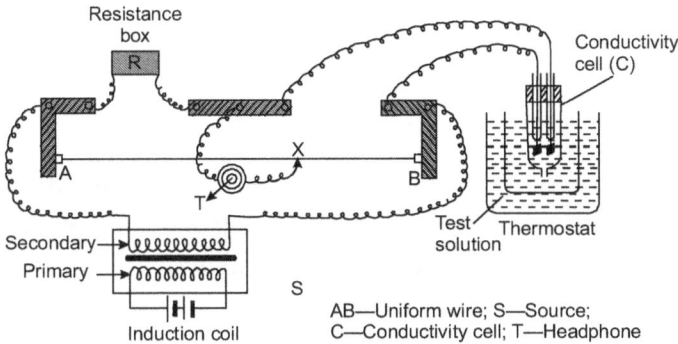

Figure 1.1   Schematic representation of Wheatstone bridge

In Figure 1.1, AB is a uniform wire and a sliding contact point X moves over it. S represents the source of alternating current connected into the circuit. C is the conductivity cell containing the solution whose resistance is to be measured. T is the headphone to detect the flow of current. When current is flowing, a known resistance R is introduced through the resistance box. The resistance should be of about the same order as that of the solution under study. The sliding contact X is then moved along the wire AB until a point of minimum sound in the headphone is detected. At this stage,

$$\frac{\text{Resistance of the solution}}{\text{Known resistance}} = \frac{\text{Length XB}}{\text{Length AX}}$$

From the above representation, if the lengths AX, XB and R are known, the resistance of the solution can be calculated. The reciprocal of the resistance gives the conductance of the solution.

The specific conductance and conductance are related as

$$\kappa = \left(\frac{l}{a}\right) \times \text{conductance}$$

where,

l = distance between the two electrodes and

a = area of cross section of the electrode.

The factor l/a is called as cell constant. The unit of cell constant is cm^{-1}. Knowing the cell constant and conductance, the specific conductance can be calculated.

Specific conductance = cell constant × conductance

The most convenient method of calculating a cell constant is by measuring the conductance of a standard solution whose specific conductance is known. The cell constant is not calculated from the values of l and a because these are difficult to be measured for a given cell. A standard solution of KCl is used whose conductivity is known at different concentrations and temperatures. The specific conductance of KCl solutions at different temperatures are given in Table 1.1.

Table 1.1 Specific conductance of potassium chloride solution at different temperatures

Concentration (N)	Specific conductance (ohm^{-1} cm^{-1})		
	0°C	18°C	25°C
1.0	0.06543	0.09820	0.11173
0.1	0.007154	0.011192	0.012886
0.01	0.0007751	0.0012227	0.0014114

$$\text{Cell constant} = \frac{\text{Specific conductance}}{\text{Measured conductance}}$$

Once the cell constant has been determined, specific conductance of any solution can be easily obtained from its measured conductance.

Types of Conductivity Cells

Dip cells It is the simplest and most commonly used conductivity cell. It consists of two electrodes (Figure 1.2) placed at a fixed distance apart and surrounded by a glass or plastic sleeve to protect the electrodes. Stirring may be required to prevent polarization of certain charged species at the electrodes.

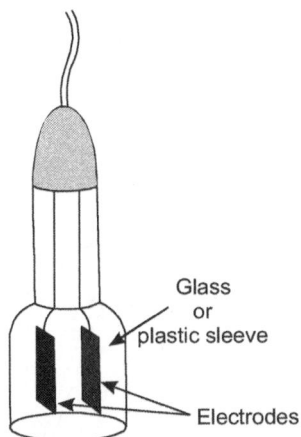

Figure 1.2   Dip cell

Flow-through cells It consists of a glass tubing in which disc electrodes are connected at fixed points. Ion chromatographic detectors are commonly of the flow-through type.

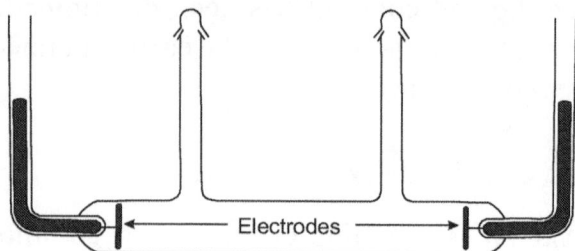

Figure 1.3   Jones cell for moderate conductivity

Jones cell   This cell consists of two separate borosilicate glass bulbs containing platinum disc electrodes (Figure 1.3). The glass bulbs are then connected by a section of glass tubing of a specified diameter and length so as to give the desired cell constant. This cell is used for more accurate measurements than dip and flow-through cells.

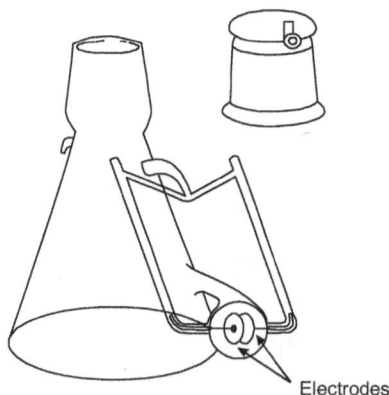

Figure 1.4   Daggett cell for low conductivity

Daggett or Kraus cells   It consists of an Erlenmeyer flask to which is attached a glass bulb containing the electrodes (Figure 1.4). These cells are usually used for low conductivities because the cell constant is small due to the size and juxtaposition of electrodes.

Pipette cells   This cell is similar to a dip cell except that the electrodes are enclosed in a pipette instead of a sleeve. It is used for extremely small volumes of sample (~2 ml).

CONDUCTIVITY MEASUREMENTS

Electrolytic conductivity is determined by two methods. The most common method uses electrodes immersed in a solution. The other method does not use electrodes and is called electrodeless conductivity measurement.

Immersed-Electrode Measurements

An electrolytic solution is placed in a cell and a current is passed through the solution. Two electrodes in contact with this solution allow the measurement of resistance. The types of electrodes and the methods for determining resistance vary with the type of current used.

AC conductivity determination   The common method of electrolytic determination uses alternating current at frequencies below 5000 Hz. Conductivity cells of AC measurements operate with two, three or four electrodes. In the two-electrode cells (Figure 1.5), the electrodes are used both for the passage of current and for the measurement of resistance. In three- and four-electrode cells, a pair of electrodes is used to measure the resistance while one or two additional electrodes are used to pass the current.

The three- and four-electrode cells allow a lower amount of current to be used, which reduces the common problem of polarization and fouling of electrodes. The electrodes used in AC cells are non-reactive materials such as platinum, titanium, solid-plated nickel, tungsten or graphite.

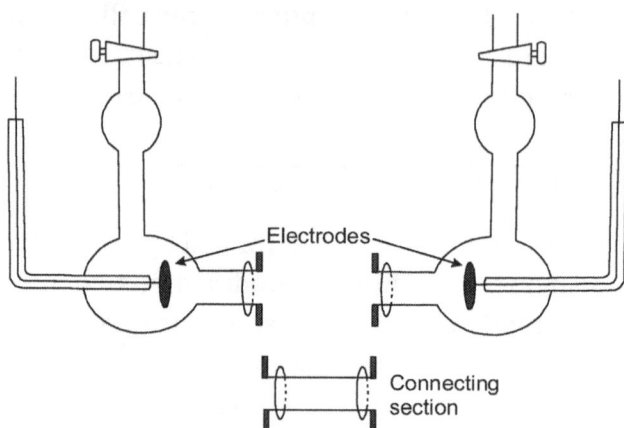

Figure 1.5   Absolute cell for AC measurement

The standard method of AC conductivity measurements uses a Wheatstone bridge to measure the resistance of the electrolyte solution in the cell. A frequency generator is used to supply an AC signal, normally in the range of 1000 to 5000 Hz to the bridge. The cell is placed in one arm of the bridge and a variable resistor is placed in the other arm. A variable capacitor in parallel to a variable resistor is used to compensate for the impedance of the cell. A differential input preamplifier, lined amplifier, or oscilloscope may be used as a null detector. At the point where a null detector shows balance, the amplitude of the potentials on each side of the null detector are equal and in phase, and the impedance of the cell is equal to the impedance of the capacitor–resistor set-up. The resistance of the cell is determined by the resistance of the variable resistor. (Because the resistance of the variable resistor is equal to the sum of the resistance of the electrolyte solution and of the leads, any resistance given by the lead is subtracted from the resistance given by the bridge.) The resistance is measured at several frequencies and extrapolated to infinite frequency to eliminate polarization effects. The commercial conductivity bridges measure the resistance at only one frequency.

DC conductivity determination In this method, the measurement can be accomplished by using direct current. It has been used to determine the electrolytic conductivity of primary standard solutions. The DC measurements are limited to cells of four-electrode type. It also needs that all four electrodes must be reversible to the electrolyte solution (Figure 1.6). This is necessary to eliminate polarization effects.

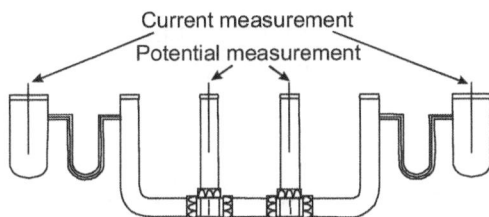

Figure 1.6   Absolute cell for DC measurement

It is much simpler than AC conductivity measurements and operates on the principle of Ohm's law. A standard resistor is placed in series with the conductivity cell. The current passing through the cell is equal to the current passing through the standard resistor. A voltmeter is used to determine the potential difference across the standard resistor. The potential difference between the two electrodes placed in the solution of interest is measured, and the resistance between the electrodes is then calculated using Ohm's law.

Electrodeless Measurements

There are two basic methods, namely inductive coupling and capacitive coupling for determining the electrolytic conductivity without placing electrodes in direct contact with the solution.

Inductive measurements use a pair of toroids to measure conductivity at frequencies around 20 MHz. An oscillator energizes the first toroid (acts as a transformer) and induces a current in a loop of solution. The solution loops through and energizes the second toroid. As the conductivity of the solution increases, the current induced in the second toroid increases. When the circuit is optimized, the current is linear with the conductivity of the solution.

Measurements using capacitive coupling require frequencies from 1 MHz to 10 MHz. Electrodes are placed outside of a glass cell. The current flowing through the cell is proportional to the resistance of the solution in the cell, thus allowing for the measurement of electrolytic conductivity.

APPLICATIONS OF CONDUCTIVITY MEASUREMENTS

Determination of Ionic Product of Water (K_w)

Water is slightly dissociated as

$$H_2O \rightleftharpoons H^+ + OH^-$$

The dissociation constant (K) of H_2O is given by

$$K = \frac{[H^+][OH^-]}{[H_2O]}$$

The concentration of ionized water may be taken as a constant since water dissociates very slightly.

$$[H^+][OH^-] = \text{constant} = K_w$$

where K_w is the ionic product of water, which is constant at a given temperature. The specific conductance of pure water at

25°C is 5.54×10^{-8} ohm^{-1} cm^{-1}. The conductance of one litre of water, therefore is 5.54×10^{-5} ohm^{-1} cm^2. The equivalent conductance of water, if it is completely ionized to give one gram equivalent of H$^+$ and one gram equivalent of OH$^-$, is taken as the sum of ionic conductance of H$^+$ and OH$^-$. It is 548.3 ohm^{-1}cm^2 equivalent^{-1}. As the conductance is proportional to the number of ions, the number of gram equivalents of H$^+$ ions per litre will be

$$\frac{5.54 \times 10^{-5}}{548.3} = 1.011 \times 10^{-7}$$

The number of gram equivalents of OH$^-$ ions will also be the same. Hence

$$K_w = [H^+][OH^-] = 1.011 \times 10^{-7} \times 1.011 \times 10^{-7}$$
$$= 1.02 \times 10^{-14}$$

Determination of Solubility of Sparingly Soluble Salts

The salts like AgX, $BaSO_4$, PbO_2, etc. are sparingly soluble in water. Their solubility cannot be determined by any chemical method. It is possible to determine even extremely small solubilities by conductance measurements.

It is required to find out the solubility of AgCl in water at 25°C. The salt is repeatedly washed, suspended in water, warmed and cooled to 25°C. Very minute quantity will be dissolved and its conductance can be determined by the usual method. The conductance of water used in the preparation of solution is also determined. The difference between the two multiplied by the cell constant will be the specific conductance of AgCl.

The solubility of AgCl is say S gram equivalent. The volume of solution containing 1 gram equivalent of AgCl will be 1000/S ml. Hence, the equivalent conductance will be $\dfrac{\kappa \times 1000}{S}$ ohm^{-1}cm^2 equivalent^{-1}.

The equivalent conductance of AgCl can be taken as the sum of ionic conductances of Ag$^+$ and Cl$^-$. So, knowing the equivalent conductance and the specific conductance of AgCl, its solubility can be easily calculated.

Conductometric Titrations

Conductance measurements are employed to find out the end points of acid–alkali and other titrations. The principle involved is that electrical conductance depends on the number and mobility of ions. Conductivities are followed during the course of titration, and values are plotted against the number of cc of the titrant added. It is not necessary that the actual specific conductivity of solution be used. Instead, any quantity proportional to it can be used in its place. Conductometric titrations have several advantages—it can be used for coloured solutions where no specific indicator is found to be satisfactory and for very dilute solutions. In order to get accurate results, the change of volume during the titration should be as small as possible. Hence, titrant should be 5 to 10 times stronger than the solution taken in a conductivity vessel.

Titration of strong acid (HCl) against strong base (NaOH). The acid is taken in the conductivity vessel and the base in the burette. The conductance of acid is due to the presence of H$^+$ and Cl$^-$. As base is added, the hydrogen ions are replaced by the slow-moving Na$^+$ ions. This is represented by

$$H^+ + Cl^- + Na^+ + OH^- \longrightarrow NaCl + H_2O$$

Hence, on continued addition of NaOH, the conductance decreases until the acid is completely neutralized. Further addition of base will result in an increase in conductance due to the presence of fast-moving hydroxyl ions. The conductance is plotted against the volume of base added. The point of intersection of two lines gives the end point, which is shown in the titration curve below.

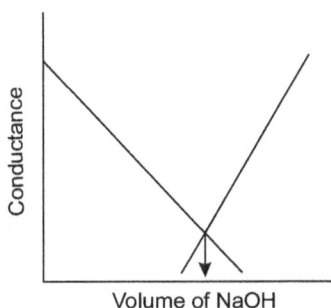

Volume of NaOH

Titration of strong acid (HCl) against weak base (NH$_4$OH). When NH$_4$OH is added to neutralize HCl, the conductance decreases at first due to the replacement of fast-moving H$^+$ ions by slow-moving NH$_4^+$ ions.

$$H^+ + Cl^- + NH_4^+ + OH^- \longrightarrow NH_4Cl + H_2O$$

After neutralization of acid, further addition of weakly ionized NH$_4$OH will not cause any appreciable change in the conductance. The titration curve is shown below.

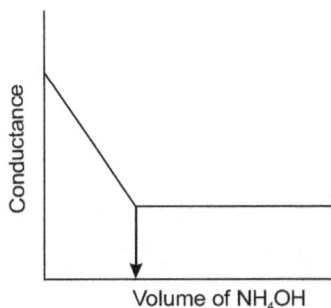

Volume of NH$_4$OH

Titration of weak acid (CH_3COOH) against strong base (NaOH)   When NaOH is added to CH_3COOH, the conductivity decreases, but since the concentration of H^+ ions in acetic acid is low, the conductance of the solution soon increases due to the formation of Na^+ and CH_3COO^- ions.

$$CH_3COOH + Na^+ + OH^- \longrightarrow CH_3COO^- + Na^+ + H_2O$$

After neutralization of acetic acid, the further addition of base, increases the conductance more sharply. The intersection of lines obtained by plotting conductance against volume of base gives the end point.

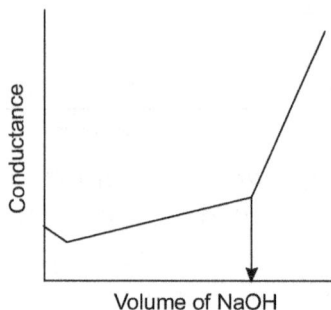

Volume of NaOH

Titration of weak acid (CH_3COOH) against weak base (NH$_4$OH)   In 1960, Gaslini and Neahm have developed an indirect method for titrating weak acids and bases. In the case of weak acid, an excess amount of NH_4OH is added to CH_3COOH to convert it into a highly dissociated ammonia salt. The ammonium ion is then titrated with lithium hydroxide giving a much more distinct end point that would be obtained for a direct titration.

In the case of a weak base, an excess of acetic acid is added to NH_4OH. The highly dissociated salt will be formed. The unreacted acetic acid is titrated with trichloroacetic acid.

Volume of NH₄OH

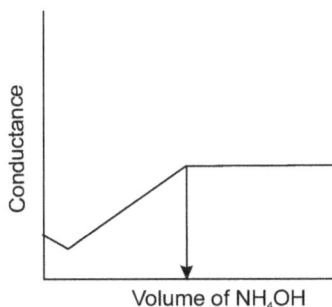

Titration of mixture of acids against weak base Let us consider a titration of mixture of HCl and CH_3COOH against NaOH. HCl being a strong acid, will get titrated first. The titration of acetic acid will commence only when HCl is completely neutralized. The titration curve is shown below. The end point A corresponds to the neutralization of HCl while B corresponds to the neutralization of CH_3COOH.

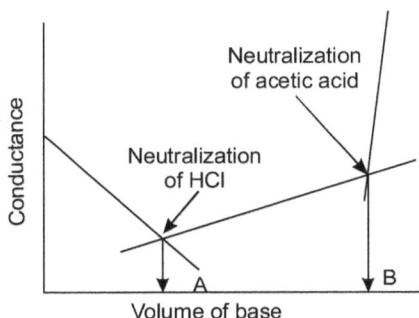

Conductometric precipitation titration The precipitation of $AgNO_3$ against KCl is carried out by conductometric methods. The precipitation reaction is represented as

$$Ag^+ + NO_3^- + K^+ + Cl^- \rightarrow AgCl^- K^+ + NO_3^-$$

When AgCl is precipitated, one salt is replaced by equivalent quantity of another salt (i.e., KCl by KNO_3).

So the conductance remains constant. After the end point, the addition of KCl increases the conductance.

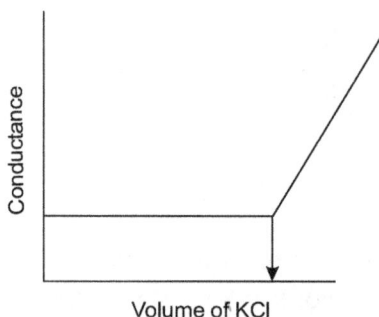

Volume of KCl

Conductometric displacement titrations   Salts of strong acid and weak base is titrated conductometrically against a strong base. Similarly salts of strong base and weak acid can be titrated against strong acid. Let us consider a titration of sodium acetate (salt of strong base and weak acid) against a strong acid, HCl.

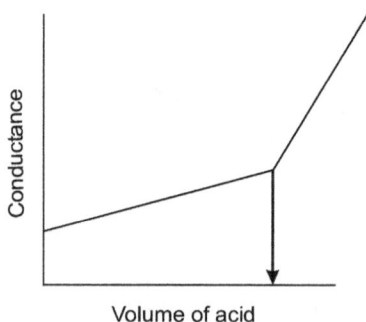

Volume of acid

The displacement reaction is represented as

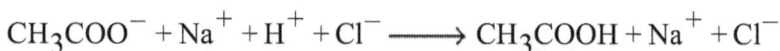

$$CH_3COO^- + Na^+ + H^+ + Cl^- \longrightarrow CH_3COOH + Na^+ + Cl^-$$

As the mobility of chloride ion is slightly higher than the acetate ion, the conductance increases slowly till equivalence point.

After the end point, further addition of HCl produces a sharp increase in conductance. The end point is determined by the intersection of two lines as shown in the curve.

In the titration of ammonium chloride (salt of weak base and strong acid) against strong acid, the conductivity may decrease or remain constant. In general, therefore, the change in conductance is very small before the end point.

Conductometric redox titrations   Most oxidation–reduction reactions involve a decrease in the hydrogen ion concentration. For example,

$$6Fe^{2+} + Cr_2O_7^{2-} + 14H^+ \longrightarrow 6Fe^{3+} + 2Cr^{3+} + 7H_2O$$

Because of the high mobility of hydrogen ions, a sharp decrease in conductance can be seen during the initial part of titration and thus we have a satisfactory conductometric method. It has been observed that, if the acidity is high in the initial stage, the relative change in hydrogen ion concentration is very small to give good precision. And also, the acidity should not be too low to permit hydrolytic precipitations. The oxidation of ferrous by dichromate is given in the titration curve below.

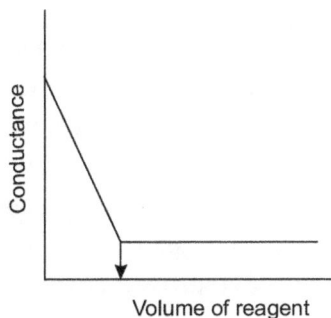

Volume of reagent

Determination of Purity of Water

The purity of water is essential for many industries. Conductivity detectors are being used to monitor the concentration of bleach in textile industry, to monitor the cleanliness of pipelines and processing systems in food industry, and to detect the spills of costly chemicals in pulp and paper industry. The conductivity of purest water (free from CO_2) has a theoretical value of 0.055 μS/cm Atmospheric CO_2 dissolved in water increases the conductivity to 0.8–1.2 μS/cm Traces of mineral impurities such as Ca or Mg may increase the conductivity to several μS/cm. The purity of water is listed with the help of conductivity detectors by measuring its resistivity.

Concentration Determination

In a solution containing only one species, the concentration of the species can be determined by knowing its conductivity. The concentration of a given species in a mixed electrolyte cannot be measured because the conductivity is non-specific. However, the total ionic content of a solution is approximated as total dissolved solid (TDS) or total ionic salt (TIS) or salinity.

Ion Chromatography

Conductivity detectors are used in chromatographic equipment to determine the elution of charged analytes. Special microconductivity cells have been developed of a flow-through pattern and placed in a thermostated enclosure. A typical cell may contain a volume of about 15 μl and may have a cell constant of ~0.15 cm^{-1}. The sensitivity is improved by the use of a bipolar square-wave pulsed current which reduces the polarization and capacitance effects, and the changes in conductivity caused by the heating effect of the current.

Problems

1. A 0.01N KCl solution shows resistance 225 ohms in a conductivity cell. The specific conductivity of 0.01N KCl solution at the temperature of experiment is 0.00141 mho/cm. If a 0.02N solution of an acid shows a resistance of 80 ohms in the same cell, find the specific and equivalent conductance of the acid.

Solution

For 0.01N solution,

Cell constant	=	Specific conductance × Resistance
	=	0.0014.1 mho/cm × 225 ohm
	=	0.3173 cm^{-1}
∴Sp. conductance of 0.02N acid solution	=	Cell constant/Resistance
	=	$\dfrac{0.3173}{80}$
	=	3.966 × 10^{-3} mho/cm
∴Eq. conductance of 0.02N acid solution	=	1000 K/N
	=	$\dfrac{1000 × 3.966 × 10^{-3}}{0.02}$
	=	198.28 mho cm^2 eq^{-1}.

2. The equivalent conductance of 1.028 × 10^{-3}N acetic acid is 48.15 mho cm^2 eq^{-1} at 298 K. Its equivalent conductance at

infinite dilution is 390.6 mho cm^2 eq^{-1}. Calculate the degree of dissociation of acetic acid.

Solution

Degree of dissociation of acetic acid,

$$\alpha = \frac{\Lambda_{eq}^{c} \ (CH_3COOH)}{\Lambda_{eq}^{\infty} \ (CH_3COOH)}$$

$$= \frac{48.15}{390.6}$$

$$= 0.1233 \text{ or } 12.33\%$$

3. A conductor cell has two parallel plates of 1.25 cm^2 area placed at 10.50 cm apart. When filled with a solution of an electrolyte, the resistance was found to be 2.0 × 10^3 ohms. Calculate the cell constant and the specific conductance of the solution.

4. Calculate the molar conductance of ammonium hydroxide at infinite dilution from the given equivalent conductance of sodium hydroxide, sodium chloride and ammonium chloride as 247 × 10^{-4}, 126.45 × 10^{-4} and 149 × 10^{-4} mho mol^{-1} m^2 at 25°C.

5. Calculate the solubility of lead sulphate in g/L from the following data: the resistance of a saturated solution of lead sulphate in water at 25°C and of water are 1850 and 11,110 ohms respectively. When measured using a conductance cell fitted with platinum electrodes of area of cross section 2 × 10^{-4} m^2 at a distance of 0.2 × 10^{-2} metre, the equivalent conductance at infinite dilution of lead nitrate, sulphuric acid and nitric acid are 150 × 10^{-4}, 460 × 10^{-4} and 450 × 10^{-4} mho m^2 eq^{-1} respectively.

6. The ionic conductance of Ag$^+$ and NO$_3^-$ ions at infinite dilution at 18°C are 55.7 and 60.8 mho/cm respectively. If the specific conductance of a decinormal AgNO$_3$ solution

at 18°C is 0.00947 mho/cm, calculate the percentage of the salt at this dilution.

7. The specific conductivity of saturated solution of silver chloride is 1.24×10^{-6} mho/cm. The mobilities of Ag^+ and Cl^- ions are 53.8 and 65.3 mhos eq^{-1} cm^2. Calculate the solubility of AgCl in g/L. The atomic weights of Ag and chloride are 108 and 35.5 respectively.

8. The conductivity of a 0.1 M KCl solution at 25°C is 0.013 mho/cm.

   i. Calculate the solution resistance between two parallel planar platinum electrodes of 0.1 cm^2 area placed 3 cm apart in this solution.

   ii. A reference electrode with a luggin capillary is placed in the following distances from a planar platinum electrode ($A = 0.1$ cm^2) in 0.1 M KCl: 0.05, 0.1, 0.5, 1.0 cm. What is R_u in each case?

References

Harned, H.S. and Owen, B.B. (1958). *The Physical Chemistry of Electrolyte Solutions*, 3rd edn. Reinhold, New York.

Light, T.S. and Ewing, W. (1990). "Measurement of electrolytic conductance." In: *Analytical Instrumentation Handbook*. Ewing, G.W. (ed.). Marcel Dekker, New York.

Loveland, J.W. (1986). "Conductance and oscillometry." In: *Instrumental Analysis*, 2nd edn. Christian, G.D. and O'Reily, J.E. (eds.). Allyn & Bacon, Boston.

Wu, Y. C. *et al.* (1987). "Review of Electrolytic Conductance Standard." *Journal of Solution Chemistry*. 16, No. 12, 985.

2

POTENTIOMETRY

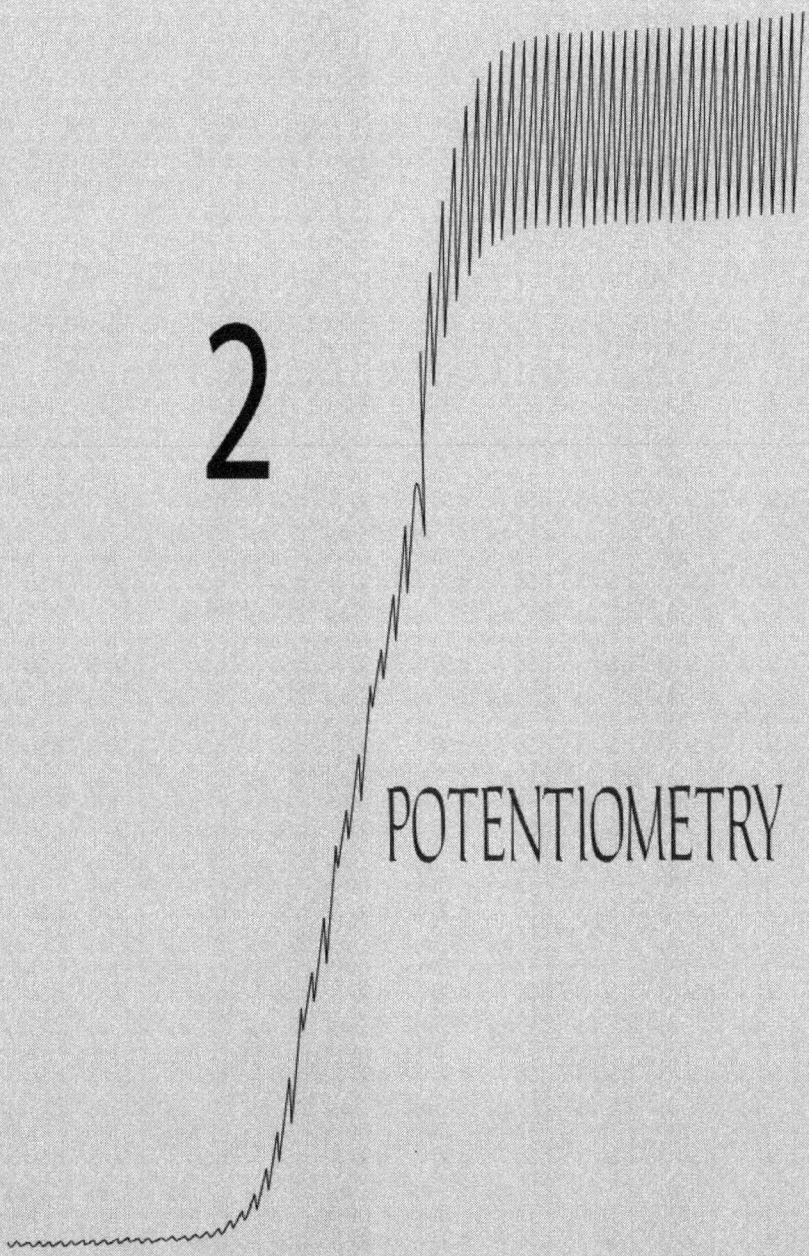

Potentiometry can be described as the measurement of a potential in an electrochemical cell. It is the only electrochemical technique that directly measures a thermodynamic equilibrium potential and in which no net current flows. Nernst (1864–1941) was instrumental in deriving the thermodynamic equilibrium relationship between the galvanic cell potential and the activity of an ion in solution. One of the unique features of potentiometry is the ability to monitor the activity of an ion in a sample rather than the concentration. The determination of activity results from the thermodynamic equilibrium relationship between the activity of an ion and the potential of a cell. Potentiometry thus is one of the few methods available for distinguishing between free (ionized) and bound complex ions in a sample as well as between the activities of different oxidation states of a given ion that may be present. A calibration plot is constructed for the determination of an unknown activity by measuring the cell potential (E_{cell}) at various known activities and graphing the results. The two electrodes used in potentiometry are the reference and indicator (working) electrodes.

Potentiometry as an analytical technique was limited to the pH determination and redox electrodes. It is useful in titration for the determination of electroactive species. Later, the development of ion-selective electrodes (ISEs) (selective for a given ion), was achieved and through the development of these electrodes, older, expensive and time-consuming analytical techniques were replaced.

PRINCIPLE

Potentiometric analysis involves measuring the potential difference (E_{cell}) between an indicator and a reference electrode.

External reference electrode		Aqueous sample	Ion-selective membrane	Internal filling solution	Internal reference electrode

$$E_{r2} \quad E_j \qquad E_{b1}\!\leftarrow\!E_d\!\longrightarrow\!E_{b2} \qquad E_{r1}$$

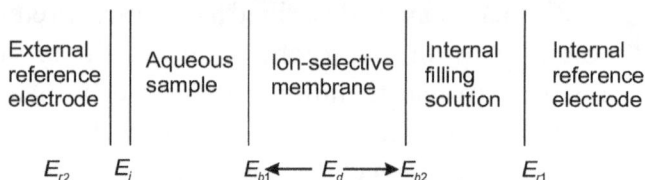

The E_{cell} of the above cell is given as

$$E_{cell} = (E_{b1} + E_d + E_{b2} + E_{r1}) - (E_{r2} + E_j)$$

The internal reference potential of the indicator electrode (E_{r1}) and the potential of the reference electrode (E_{r2}) are fixed. It is a constant for a given electrode system at a fixed temperature. The junction potential, E_j, results from the different mobilities of individual ions and the separation of charges that results across the junction, known as a salt bridge, of two solutions of varying compositions. Ions that have higher mobilities diffuse across the interface ahead of slower moving ions, resulting in this separation of charges. An equitransferent electrolyte in which the cations and anions are approximately equal in mobility (such as KCl and KNO_3) minimizes E_j. The E_{cell} can be written as follows, by considering E_{r1}, E_{r2} and E_j as constant.

$$E_{cell} = E_{b1} + E_d + E_{b2} + K$$

where K is a constant.

$$E_m = E_{b1} + E_d + E_{b2}$$

where, E_{b1} is the phase boundary potential at the external surface of the membrane, which is created by the separation of charges resulting from the partitioning of a given ion into the membrane. The potential E_{b2} arises in a similar way on the inner surface of the indicator electrode membrane. The potential E_d arises from the separation of charges due to the diffusion of ions through the membrane. Generally, E_d is small on the time scale of potentiometric measurements and therefore it can be ignored. The potential E_{b2} can be made constant by fixing the concentration of the ions in contact with the inner membrane. Hence, the entire E_{cell} can be directly related to the ability of a membrane in contact with the sample solution to be permselective (allow only ions of one charge to partition into the membrane) and otherwise called as membrane potential, E_m. The selectivity of the membrane for a targeted ion depends on how specific the interaction of the membrane is with the analyte.

By using Henderson approximation, a thermodynamic relationship is acquired between the E_{cell} and the activity of the analyte, which assumes a linear concentration gradient of each ion across the membrane. Assuming that the membrane is only permeable to a single ion,

$$E_{cell} = K + \frac{RT}{zF} \ln (a_i)$$

where,

R is the gas constant,

K is a constant of the measurement system,

T is the temperature (in Kelvin),

z is the charge on the analyte ion,

F is Faraday's constant and,

a_i is the activity of the analyte ion i.

At 25°C and with a charge of +1, E_{cell} vs log a_i should be linear over the working range of the electrode with a slope of

59 mv/decade change in activity. Since membranes are not completely specific for a given ion, a more general relationship is described by the Nicolsky Eisenman equation. It is given by

$$E_{cell} = K + \frac{RT}{z_i F}\left(\ln a_i + K_{ij}\, a_j^{\frac{z_i}{z_j}} \right)$$

where,

a_i is the activity of the analyte ion i,

a_j is the activity of the interfering ion j,

z_i and z_j are the charges of ions i and j respectively and

K_{ij} is the selectivity coefficient.

If more than one interfering ions are present, additional $K_{ij}\, a_j^{z_i/z_j}$ terms are needed to describe the behaviour of the electrodes, as there are additional ions that partition into the membranes. The magnitude of interference in the measurement of a_i is determined by each K_{ij}.

$$E_{cell} = K + \frac{RT}{z_i F}\, \ln\left(a_i + \sum_j K_{ij} a_j^{z_i/z_j} \right)$$

If $K_{ij} = 1$, then the membrane responds equally to i and j. The smaller the K_{ij} the lesser is the interference observed from the ion j. The selectivity coefficients are determined experimentally and the values will be provided with commercially available ion-selective electrodes.

Another potential source that contributes to E_{cell} is called asymmetry potential, E_{asym}. This potential is assumed to be constant and contributes only a very small potential to the overall E_{cell}. This E_{asym} arises because of the differences that exist in the internal and external surfaces of the indicator membrane.

INSTRUMENTATION

The instrumentation required to perform potentiometric measurements includes a reference electrode, an indicator electrode, and a high-input-impedance potentiometer.

Reference Electrodes

All electrode potentials are quoted with reference to the standard hydrogen electrode (SHE) and hence, this must be regarded as the primary reference electrode. The hydrogen electrode operates in a solution containing hydrogen ions at unit activity (usually hydrochloric acid) and hydrogen gas at 1 atmosphere pressure.

Although the hydrogen electrode is the primary reference electrode, it has some disadvantages in preparing and operating satisfactorily. The platinum black coating of the electrode is susceptible to poisoning by a variety of substances and so cannot be used in the presence of oxidizing or reducing agents. To overcome these difficulties, secondary reference electrodes are used. They are (i) calomel electrodes (ii) silver–silver chloride electrodes.

Calomel electrodes The most widely used reference electrode is the calomel electrode. A calomel half-cell is one in which mercury and calomel (mercurous chloride) are covered with a 0.1 M,1M or saturated solution of potassium chloride. These electrodes are referred to as the decinormal, the molar and the saturated calomel electrodes (SCE) and have the potentials of 0.3358, 0.2824 and 0.2444 V relative to the standard hydrogen electrode at 25°C. Of these electrodes, the SCE is most commonly used because of the suppressive effect of saturated KCl on liquid junction potentials. However, this electrode suffers from the drawback that its potential varies rapidly with a change in temperature owing to changes in the solubility of potassium chloride, and restoration of a stable potential may be slow owing to the disturbance of the calomel–potassium chloride equilibrium. The potentials of decinormal and molar electrodes that are less affected by the change in temperature are to be preferred, in cases where accurate values of electrode potentials are required. The electrode reaction is given by

$$Hg_2Cl_2(s) + 2e^- \longrightarrow 2Hg\ (l) + 2Cl^-$$

Silver–silver chloride electrodes   The electrodes consist of a silver wire or silver-plated platinum wire coated electrolytically with a thin layer of silver chloride dipped into a potassium chloride solution of known concentration. Saturated potassium chloride is the most commonly used. The potential of the electrode is governed by the activity of the chloride ions and is 0.1989 V vs SHE. The silver–silver chloride electrodes should be used in solutions containing species that can precipitate or complex with silver (such as halides, proteins and sulphide).

Indicator Electrodes

There are many types of indicator electrodes available, the selection of which depends on the application desired. They are classified as

   i.   metallic electrodes

   ii. membrane electrodes

The change in the potential of the indicator electrode with respect to the fixed potential of the reference electrode is the analytical parameter observed.

Metallic electrodes   There are four types of metallic electrodes, electrodes of the first, second and third kinds, and redox electrodes. The electrodes of first kind and redox electrodes are dipped in nitric acid to clean the surface before use.

Electrodes of first kind   An electrode of first kind is a metal wire (solid plate or mesh) that can respond to its own metal cations in solution, e.g. Ag wire and Ag^+ in solution. These electrodes exhibit poor selectivity, usually responding preferentially to cations that are easily reduced. They do not respond to analyte if the metal ion has been oxidized. However, the surface of the electrode itself can be oxidized in the presence of air, making it impossible to obtain accurate measurements at low ion activities. The most commonly used electrodes are Ag/Ag^+ and Hg/Hg^+. The other metal electrodes such as Cu/Cu^{2+}, Cd/Cd^{2+}, Sn/Sn^{2+} and Pb/Pb^{2+} are used if O_2 is removed by deaeration of the sample solution.

Electrodes of second kind   An electrode of second kind consists of a metal immersed in a saturated solution containing one of its sparingly soluble salts or coated with such salt. This electrode responds to the activity of the anion, despite the absence of direct electron transfer between the anion and the electrode. The activity of free metal ion is controlled by complexation or precipitation with the anion. Thus, the observed potential can be related indirectly to the activity of the anion. Any other anion present in solution, that can complex or form an insoluble salt with a metal ion will interfere with the measurement of the activity of species under examination.

Both Ag/AgCl and SCE described in reference electrodes are the examples of electrodes of this kind. Halides and other anions that can form insoluble salts with silver or mercury or other metals may be detected with this kind of electrode.

Electrodes of third kind   Like electrodes of second kind, electrodes of third kind depend on solution equilibrium but in this case two solution equilibria are involved. These electrodes respond to a cation other than that of the electrode metal. For a better response, the complex between the metal cation and anion must be more stable than the complex between the cation under study and the anion, and there must be excess of uncomplexed cations in the solution. Example for this kind is a mercury electrode, with EDTA as the common anion that responds to calcium.

Redox indicator electrodes   Redox indicator electrodes can be used to detect redox species. The potential of this electrode depends on the ratio of activities of both species in a redox couple. The most inert metals such as Au, Pt and Pd are some good examples of this kind of electrode. A carbon surface, on which redox reactions are typically fast, can also be used. The limitation of these electrodes is that electron transfer is not always reversible resulting in non-reproducible potentials.

Membrane electrodes   Several types of membranes have been used that allow for a variety of different analytically useful ion-specific electrodes. They are discussed in the following section.

Glass electrodes   The glass electrode or pH electrode is the most widely used hydrogen-ion-responsive electrode, and its use is dependent upon the fact that when a glass membrane is immersed in a solution, a potential is developed which is a linear function of the hydrogen ion concentration of the solution.

The arrangement of a glass electrode is shown in Figure 2.1. A bulb is immersed into the solution the hydrogen ion concentration of which is required to be measured. The electric circuit is completed by filling the bulb with a solution of HCl (0.1 M) and by inserting an Ag–AgCl electrode. The concentration of HCl is maintained constant. The potential of Ag–AgCl electrode inserted into it will be constant. Hence, the potential between the HCl and the inner surface of the glass blub is constant. The only potential which can vary is that existing between the outer surface of the glass bulb and the test solution in which it is immersed, and so the overall potential of the electrode is governed by the [H$^+$] of the test solution.

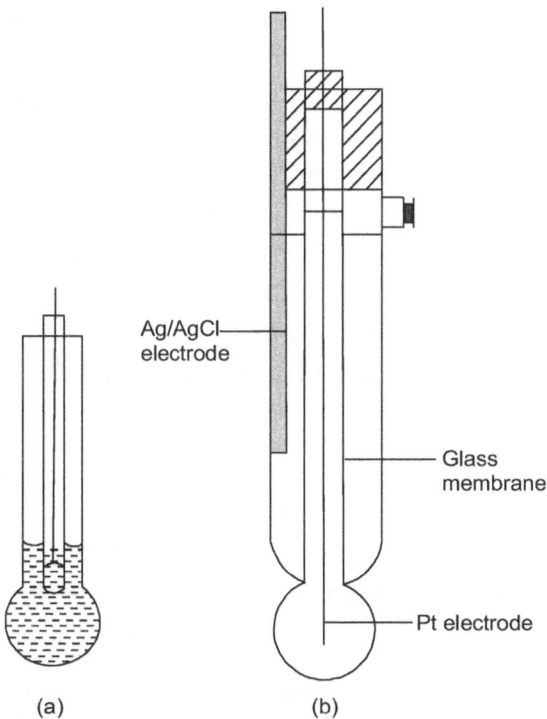

Figure 2.1   (a) Glass electrode (b) Combination electrode

Glass electrodes are now available as **combination electrodes** which contain the indicator electrode (Ag/AgCl) combined in a single unit. The glass electrode should be thoroughly washed with distilled water after each measurement and then rinsed with several portions of test solution before making the next measurement. The glass electrode should not be allowed to become dry, except during long periods of storage. It is returned to its responsive condition when immersed into distilled water for at least 12 hours prior to use.

The pH electrode may be calibrated using commercially available standards. Specific calibration instructions accompany all pH meters, but a multipoint calibration is best chosen so that the expected pH of the sample is bracketed by that of the standard buffers used. The accuracy of the measurement can be maximized by allowing sufficient time for equilibration of each sample, by using highly accurate pH buffers in the calibration, and by calibrating the pH meter at the same temperature as the sample to be analysed. In very basic pH solutions, the apparent pH is usually less than the real pH due to the response of the membrane to sodium ions. This error can be minimized by altering the glass composition. Electrodes especially designed for measurements at high pH are also available.

Additionally, glass electrodes selective for other cations have been developed using different glass compositions. Glass electrodes are commercially available for Li^+, Na^+, K^+, Rb^+, Cs^+, Ag^+, Fe^{3+}, Pb^{2+} and Cu^{2+}. Some autoclavable electrodes are also available.

Solid membrane electrodes   Solid-state electrodes can be obtained as a single crystal or as a disc pressed from finely divided crystalline material. It may be advantageous to incorporate the crystalline material into an inert carrier such as a suitable polymer, thus producing a heterogeneous membrane electrode.

The best example for a single crystal electrode is lanthanum fluoride electrode. A crystal of LaF_3 is sealed at the bottom of a plastic container to produce a fluoride ion electrode. The container is charged with a solution containing KCl and KF, and carries a silver wire which is coated with AgCl at its lower end, thus acting as a reference electrode.

The LaF_3 crystal is a conductor for fluoride ions which being small can move through the crystal from one lattice defect to another. An equilibrium is established between the crystal face inside the electrode and the internal solution. When this electrode is placed in a solution containing F^- ions under examination, an equilibrium is established at the external surface of the crystal. The activities of a fluoride ion at the two faces of the crystal are different and so a potential is established. Since the potential at the inner surface is constant, the resultant potential is proportional to the activity of the fluoride ion of the test solution.

The best example for the pressed disc (or pellet) type of crystalline membrane is silver sulphide in which Ag^+ ions can migrate. The pellet is sealed at the base of a plastic container and a contact is made by means of a silver wire with its lower end embedded in the pellet. This wire establishes an equilibrium with Ag^+ in the pellet and this functions as an internal reference electrode. When this electrode is placed in a solution containing silver or sulphide ions, it acquires a potential which depends on the activity of Ag^+ or S^{2-} ion respectively.

If the pellet contains a mixture of silver sulphide and insoluble sulphide of Cu (II), or Cd (II), or Pb (II), the electrodes respond to the activity of the appropriate metal ion in the test solution.

Ion-exchange electrodes   Ion-exchange electrodes are prepared by using an organic liquid ion-exchanger which is immiscible with water or using an ion-sensing material dissolved in an organic solvent which is immiscible in water, and placed in a tube sealed at the lower end by a thin hydrophobic membrane such as cellulose filter—aqueous solutions will not penetrate this film. The electrode set-up is shown in Figure 2.2. The membrane (A) seals the bottom of the electrode vessel which is divided by the central tube into an inner compartment (B) and reservoir or outer compartment (D). The inner compartment (B) contains an aqueous solution with a known concentration of chloride of the metal ion to be determined. This solution is also saturated with AgCl and carries a silver electrode (C), which thus functions as a reference electrode. The liquid ion-exchange material (ionophore) is placed in the outer compartment (D). The pores of the membrane become impregnated with the organic liquid, which allows contact with the aqueous test solution in which the electrode is placed.

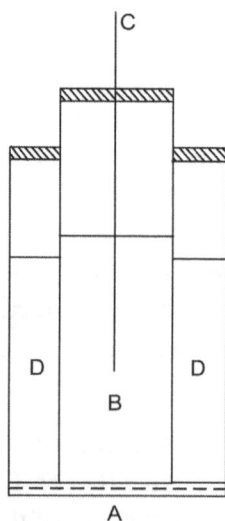

Figure 2.2   Ion-exchange electrode

It is now usual to prepare solid ion-exchange membranes by dissolving the liquid ion-exchange material together with polyvinyl chloride (PVC) in a suitable organic solvent such as tetrahydrofuran and then allowing the solvent to evaporate. A disc is cut from the flexible residue and is cemented to a PVC tube to produce an electrode vessel, in which the PVC membrane replaces the cellulose acetate and reservoir material, so that only a single compartment is needed. The ion-exchange materials or ionophores are prepared as either charged cation or anion exchanger, or as neutral carriers, selective for a given ion. The generation of potential (due to charge separation) across the membrane produces the signal observed. The selectivity of an electrode is determined by the degree to which an ionophore complexes selectively with an analyte ion. These electrodes have a limited lifetime because both the plasticizer and inophore are leached from the membrane, over time. Sensors with an increased lifetime have been developed by increasing the lipophilicities of both the plasticizer and the ionophore, and by covalently attaching the ionophores directly to the polymers.

Another configuration of polymer electrodes uses a membrane-coated metal wire. The main interest in the coated wire electrodes is in the area of miniaturization and in ion analysis.

Gas-sensing electrodes   These electrodes can be used both to monitor the gases directly and to determine the concentration of ions whose conjugated acid or base is a gaseous species. To determine the proportion of any gas in a stream of gases, the gaseous mixture is passed through the scrubber, where the gases are dissolved in water and the resultant liquid is then examined with the appropriate gas-sensing electrodes.

The ion-selective and external reference electrodes are placed behind a thin, gas-permeable membrane through which the gas diffuses. The essential features of a gas-sensing electrode are similar to that shown in Figure 2.2. The vessel (B) and its attachments are relevant to this type of electrode. The membrane A is a microporous membrane manufactured from either polytetrafluoroethylene or polypropylene, both of which are water-repellent and are not penetrated by aqueous solutions, but they allow gas molecules to pass through. Thus membrane A is permeable to the dissolved gas in the test solution. This kind of membrane is used for the determination of NH_3, CO_2 and NO_2. A glass pH electrode or any other suitable ion-selective electrode and silver–silver chloride (reference) electrode is inserted in the vessel (B) which is filled with an internal solution of sodium chloride or any electrolyte appropriate to the gas which is being determined. The electrolytes used are given in the following table. The internal solution in B contains sodium chloride and an electrolyte appropriate to the gas which is being determined.

Gases	Electrolytes
NH_3	NH_4Cl
CO_2	$NaHCO_3$
NO_2	$NaNO_2$

Biocatalytic electrodes   This electrode makes use of an enzyme to convert the substances to be determined into an ionic product which can itself be detected by a known ion-selective electrode. The best example is the urea electrode, in which the enzyme urease is employed to hydrolyse urea. The reaction is given as

$$CO(NH_2)_2 + H_2O + 2H^+ + \xrightarrow{\text{urease}} 2NH_4^+ + CO_2$$

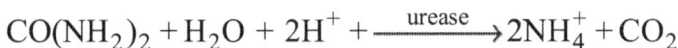

and the progress of the reaction can be followed by means of a glass electrode which is sensitive to ammonium ions. The concentration of ammonium ions determined can be related to the amount of urea present.

The high cost of an enzyme may be avoided by incorporating a small amount of enzyme in an immobile layer. The biocatalytic layer can be attached in a variety of ways, including immobilization in a gel, covalent attachment to the polymer support of the ISE, and direct absorption on to the surface of the electrode. The response time of these electrodes is longer than those of ISEs. As a result of the high selectivity of enzymes coupled with the selectivity of a membrane responsive to the product of the enzyme-catalysed reaction, very few species interfere with the biocatalytic electrodes. However, enzyme inhibitors (if present) potentially interfere with the biocatalytic electrodes. Thus, biocatalytic ISEs can also be used to monitor the concentration of inhibitors indirectly through the effect of the inhibition on the rate of the enzyme reaction.

ANALYTICAL INFORMATION

Electrodes must be calibrated before use in potentiometric analysis. The liquid junction potential (E_j) is unknown and changes with the ionic strength. Similarly the E_{asym} changes with time, making it impossible to determine K in the Nernst equation. It was assumed that K does not change during the analyte determination. However, this is not truly the case. The ionic strengths of the calibration standards and of the sample are exactly the same and the E_{asym} can change over time. Hence, frequent calibration is required.

Even though ISEs respond to activity, the concentration of ions may be determined using potentiometry. Concentration (c) is directly related to activity (a) by a proportionality constant known as activity coefficient (f).

$$a = c \times f$$

The activity coefficient is calculated using the Debye–Huckel equation.

$$\log f = -0.51z^2, \sqrt{\mu} \text{ at } 25°C$$

where, z is the charge of the ion and μ is the ionic strength of the solution. At very low concentrations of sample, the activity coefficient approaches unity. Hence, the activity and concentration of an ion are equal. As the concentration and ionic strength increases, activity coefficient decreases, resulting in activity values that deviate from the determined concentration. This deviation is much pronounced for ions with multiple charges. To minimize this deviation, the ionic strength of both the standard and sample solutions is fixed by using an ionic strength adjuster. Thus activity coefficient remains unchanged and hence activity is directly proportional to concentration.

There are two approaches for determining an unknown concentration. The first approach is a calibration curve which is obtained by plotting the E_{cell} against logarithm of the concentration for standard analyte solution. When E_{cell} of the sample is determined, the calibration curve may then be used to determine the corresponding concentration of the analyte ion. The drawback of this method is that matrix effects may result in calculated concentrations that deviate from the actual concentration of the sample. The second approach is the standard addition method. After determining E_{cell} of the sample, one or two small additions of standard solution are made to the sample and their E_{cell} is measured. The analyte concentration is again determined graphically. Minimum matrix inference is observed in this method.

ADVANTAGES

1. Ion-selective electrodes are inexpensive and simple-to-use and have an extremely wide range of applications and wide concentration range.

2. The most recent plastic-bodied solid-state or gel-filled models are very robust and durable, and ideal for use in either field or laboratory environments.

3. The measurement is rapid and easy when the ions to be determined are in low concentrations and where interfering ions are absent.

4. They are particularly useful in biological/medical applications because they measure the activity of the ion directly rather than the concentration.

5. ISEs are one of the few techniques, which can measure both positive and negative ions.

6. They are unaffected by the colour and turbidity of the sample.

7. ISEs can be used in aqueous solutions over a wide temperature range. Crystal membranes can operate in the range of 0°C to 80°C and plastic membranes in the range of 0°C to 50°C.

APPLICATIONS

Potentiometric Titrations

A potentiometric titration involves the measurement of potential of a suitable indicator electrode as a function of titrant volume. The information provided by a potentiometric titration is not the same as that obtained from a direct potentiometric measurement. For example, the direct measurement of 0.100 M solution of HCl and acetic acid would yield two different hydrogen ion concentrations because the latter is only partially dissociated. In contrast, the potentiometric titration of equal volumes of the two acids would require the same amount of standard base because both the solutes have the same number of titratable protons.

Potentiometric titrations provide the data that are more reliable than that provided by titrations that use chemical indicators, and are particularly useful with coloured or turbid solutions and for detecting the presence of unsuspected species.

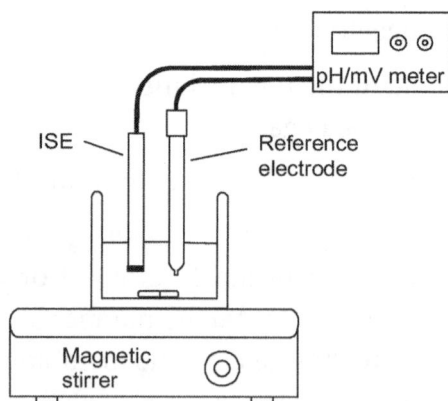

Figure 2.3   Apparatus for potentiometric titrations

Figure 2.3 illustrates a typical apparatus for performing a manual potentiometric titration. Its use involves measuring and recording the cell potential (in units of mV or pH) after each addition of a titrant. The titrant is added in larger increments at the outset and in smaller and smaller increments as the end point is approached (as indicated by larger changes in response per unit volume).

End point detection   Several methods can be used to determine the end point of a potentiometric titration. The most straightforward approach involves a direct plot of the potential as a function of reagent volume. In Figure 2.4a, the midpoint in the steeply rising portion of the curve is estimated visually and taken as the end point.

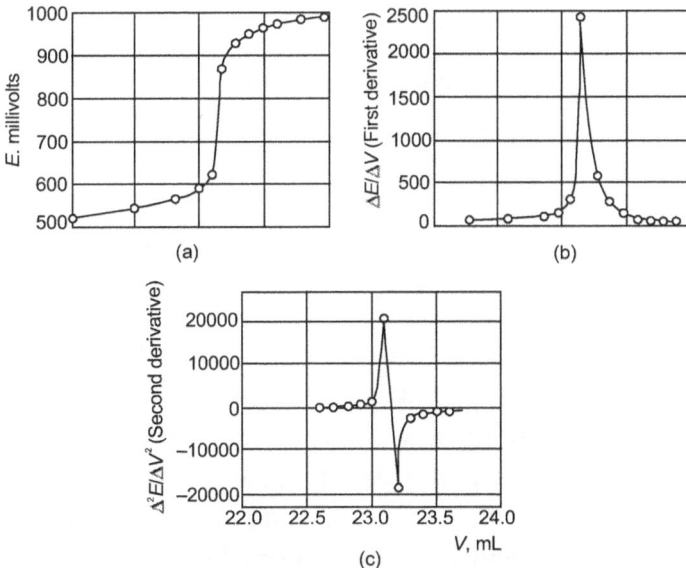

Figure 2.4   End point detection (a) plot showing potential (*E*) as a function of reagent volume (b) plot showing $\Delta E/\Delta V$ as a function of average volume (c) plot showing $\Delta E^2/\Delta V^2$ as a function of average volume

A second approach to the end point detection is to plot the change in potential per unit volume of titrant, i.e., $\Delta E/\Delta V$ as a function of the average volume V. As shown in Figure 2.4b, a curve is obtained with a maximum that corresponds to the end point.

A third approach to the end point is to plot the second derivative of potential with respect to volume as a function of the average volume Figure 2.4c. The titration data changes sign at the end point and this change is used as the analytical signal in some automatic titrators.

Chemical reactions   The majority of potentiometric titrations involve chemical reactions which can be classified as

  i. neutralization reactions

  ii. oxidation–reduction reactions

  iii. precipitation reactions

  iv. complexation reactions

For each of these different types of reactions, certain general principles can be enunciated.

Neutralization reactions   The indicator electrode may be a hydrogen, glass or antimony electrode; a calomel electrode is generally employed as the reference electrode. A known volume of the acid to be titrated is placed in a beaker provided with a magnetic stirrer. The indicator electrode is dipped into the test solution and a saturated calomel electrode is connected through a salt bridge. The cell thus formed is connected to a potentiometer and its emf is measured. A standard solution of base taken in the burette is added in small increments. After each addition of the base, the solution is stirred and the emf of the cell is measured. When the emf begins to show a tendency to increase appreciably, the increments are made smaller and smaller until near the end point.

A complete cell is represented as

Pt/0.1HCl/glass/expertl.soln//KCl,AgCl/Ag

The emf of the cell is determined potentiometrically.

$$E_{cell} = E_{right} - E_{left}$$
$$E_{cell} = 0.2444 - E_G^\circ - 0.059 \log [H^+]$$
$$= 0.2444 - E_G^\circ + 0.059 \text{ pH}$$
$$E_{cell} - 0.2444 + E_G^\circ = 0.059 \text{ pH}$$

where, E_G° is the standard glass electrode potential.

The change in the electrode potential or in the emf of the cell and in the standard reference electrode is thus proportional to the change in the pH during titration. Hence the strength of an acid can be determined. Satisfactory results are obtained in all cases except

i. those in which either acid or alkali is very weak and the solutions are dilute and

ii. those in which both the acid and the base are weak.

The method may be used to titrate a mixture of two acids that differ greatly in their strengths. The first break in titration curve occurs when the stronger of the two acids is neutralized, and the second occurs when neutralization is complete. This method is successful only when the two acids or bases differ in strength by at least 10^5 to 1.

Redox reactions   The potentiometric study of an oxidation–reduction reaction is similar to that of the neutralization reaction. Consider the titration of ferrous sulphate against dichromate solution.

$$Fe^{3+} + e^- \longrightarrow Fe^{2+}$$

The potential (E) acquired by the indicator electrode at 25°C is

$$E = E^0 + \frac{0.059}{n} \log \frac{Fe^{2+}}{Fe^{3+}}$$

where E^0 is the standard potential of the system. The potential of the immersed electrode is thus controlled by the ratio of the concentrations. During the reduction of oxidizing agent, the ratio, and therefore the potential changes more rapidly in the vicinity of the end point of the reaction. The titration curve is characterized by a sudden change of potential at the equivalence point. The indicator electrode is usually a platinum wire or foil, and the oxidizing agent is generally taken in the burette.

Precipitation reactions   The ion concentration of the precipitation reaction at equivalent point is determined by the solubility product of a sparingly soluble material formed during the titration. Consider the precipitation of $AgNO_3$ by a standard solution of KCl. An indicator electrode (either Ag wire or platinum wire plated with Ag) is placed in a solution of Ag^+ and is connected to a calomel electrode through a salt bridge.

$$Ag/AgNO_3 \text{ (aq)} \parallel KCl, Hg_2Cl_2(s), Hg$$

A standard solution of KCl is added in small amounts from a burette. As the titration proceeds, the silver ions get gradually precipitated as AgCl. The concentration of Ag^+ goes on decreasing and hence, the potential of the indicator electrode changes. This change is very rapid at the equivalence point, because the concentration of Ag^+ ions becomes very small because of slight solubility of AgCl.

One must keep in mind that, since a halide is to be determined, the salt bridge must be a saturated solution of potassium nitrate. In many cases, the use of appropriate ion-selective electrodes will be possible.

Complexation reaction Complex formation results from the interaction of a sparingly soluble precipitate with an excess of reagent. For example, when potassium cyanide is titrated against silver nitrate, the silver cyanide (initially produced) dissolves in excess KCN to give the complex ion $[Ag(CN)_2]^-$. This complexation continues up to the point where all the cyanide ions have been converted to the complex ion, the increasing concentration of which also means a gradually increasing concentration of free silver ions and hence a gradual rise in the potential of the silver electrode. Towards the end point, there is a marked rise in the potential. If $AgNO_3$ is added after the end point, the emf changes very slowly and silver cyanide is precipitated. The silver electrode (indicator electrode) and the calomel electrode (reference electrode) are separated by a salt bridge containing KNO_3 or K_2SO_4 as an equitransferent electrolyte.

For complexation titrations involving EDTA, mercury indicator electrode is used. A mercury electrode is inserted in a solution containing the metal ions to be estimated and a small added quantity of mercury (II)–EDTA complex ($[HgY]^{2-}$, where Y = EDTA). The half-cell is represented as

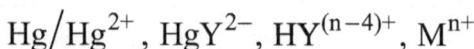

$$Hg/Hg^{2+}, HgY^{2-}, HY^{(n-4)+}, M^{n+}$$

The potential of this half-cell is given by

$$E = E^0_{Hg^{2+}/Hg} + \frac{2.303\,RT}{2\,F} \log \frac{[HgY^{2-}]}{\left[HY^{(n-4)+}\right]} \cdot \frac{K_{MY}}{K_{HgY}} + \frac{2.303\,RT}{2\,F} \log[M^{n+}]$$

where, K_{MY} and K_{HgY} are the stability constants of the metal–EDTA and Hg–EDTA complexes. The first two terms on the right side of the equation remain constant. Hence, the potential of the mercury electrode depends on the concentration of the metal ion to be determined. In complexometric titrations, the experimental condition, viz. pH must be taken care of. At a lower pH (say below 2), the Hg–EDTA complex dissociates and at a higher pH (say above 11), the oxygen reacts with mercury leading to a distorted titration curve. Hence, pH must be maintained with the help of buffer during the course of the titration.

Other Applications

Ion-selective (membrane) electrodes are used in a wide variety of applications for determining the concentrations of various ions in Ag solutions. ISEs have been used in various fields which are listed below.

Table 2.1   Applications of ion-selective electrodes

Field	Ions determined	Applications
Pollution monitoring	CN, F, S, Cl, NO_3, etc.	In effluents and natural waters.
Agriculture	NO_3, Cl, NH_4, K, Ca, I, CN, etc.	In soils, plant material, fertilizers and feedstuff.
Food processing	NO_2, NO_3	Meat preservatives.
	Salt content	Meat, fish, dairy products, fruit juices, brewing solutions.
	F	In drinking water and other drinks.
	Ca	In dairy products and beer.
	K	In fruit juices and wine making.
	NO_3	Corrosive effect of NO_3 in canned foods.
Detergent manufacture	Ca, Ba, F	In studying effects on water quality.
Paper manufacture	S, Cl	In pulping and recovery of cycle liquors.
Explosives	F, Cl, NO_3	In explosive materials and combustion products.
Electroplating	F and Cl	In etching baths
	S	In anodizing baths
Biomedical laboratories	Ca, K, Cl	In body fluids (blood, plasma, serum, sweat).
	F	In skeletal and dental studies.
Education and research		Wide range of applications.

Problems

1. The potential of the cell, $Pt/H_2/H_2SO_4/Hg_2SO_4/Hg$ at 25°C is 0.61201 in 4 M H_2SO_4. Calculate the mean ionic activity coefficient in 4 M H_2SO_4 (Given $E^0 = 0.91515$ V).

   Solution

   At LHS electrode: $H_2 \rightleftharpoons 2H^+ + 2e^-$

   At RHS electrode: $Hg_2SO_4 + 2e^- \rightleftharpoons 2Hg + SO_4^{2-}$

   Net reaction: $H_2 + Hg_2SO_4 \rightleftharpoons 2H^+ + 2Hg + SO_4^{2-}$

   $$E = (E^0_{H_2} - E^0_{Hg/Hg_2SO_4}) - RT/2F \ln a_{H_2SO_4}$$

   $$0.61201 = 0.91515 - 0.0591/2 \log a_{H_2SO_4}$$

   $$a_{H_2SO_4} = 1.2772$$

   $$a = a \pm n$$

   $$a_\pm = a^{1/n} = 1.2772^{1/3} = 1.0850$$

   $$m_\pm = (m_+ n_+ \cdot m_- n_-)^{1/n} = (8^2 \cdot 4^1)^{1/3} = 6.3496$$

   $$g_\pm = a_\pm / m_\pm = 1.0850/6.3496 = 0.171$$

2. Device a cell for evaluating the solubility product of AgCl. Calculate the solubility product.

   Solution   Solubility product (K_{sp}) is measured by constructing concentration cell method. The following cell is constructed to measure the K_{sp} of AgCl. The cell is constructed by immersing silver electrode in 0.01 M of $AgNO_3$ and another silver electrode in potassium chloride of 0.05 M. Aliquot amount of silver nitrate is added to KCl.

   $Ag/AgCl$, KCl (0.05)//$AgNO_3$ (0.01)/Ag

The standard potential of the above cell is given as

$E = 0.059 \log C_2/C_1$

$E = 0.059 \log 0.01 + 0.059 \log [Cl^-] - 0.059 \log K_{sp}$

By measuring the emf of the cell, solubility product can be calculated.

3. Consider the cell:

$$Cu/M/Fe^{2+}, Fe^{3+}, H^+ // Cl^-/AgCl/Ag/Cu$$

Would the cell potential be independent of the identity of M (e.g. graphite, gold, platinum) as long as M is chemically inert?

4. Would Na_2H_2EDTA be a good ion exchanger for a liquid membrane electrode? How about $Na_2H_2EDTA-R$, where R designates a C_{20} alkyl substituent?

5. Comment on the feasibility of developing selective electrodes for the direct potentiometric determination of uncharged substances.

6. One often finds pH meters with direct read-out to 0.001 pH unit. Comment on the accuracy of these readings in comparison to pH from test solution. Comment on their meaning in measurements of small changes in pH in a single solution (e.g. during titration).

7. Given the half-cell of the standard hydrogen electrode.

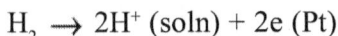

$$Pt/H_2 (a = 1)/H^+ (a = 1) \text{ (soln)}$$

$$H_2 \longrightarrow 2H^+ \text{ (soln)} + 2e \text{ (Pt)}$$

Show that the potential difference between the platinum and the solution is not zero, though the emf of the half-cell reaction is taken as zero.

References

Janata, J. (1990). "Potentiometric microsensors." *Chem.Rev.* 90: 691–703.

Koryta, J. (1990). "Theory and applications of ISEs." *Anal.Chem.Acta.* 233: 1–30.

Koryta, J. (1991). *Ions, Electrodes and Membranes,* 2nd edn. Wiley, New York.

Lewenstam, A., Maj-Zurawska, M. and Hulanicki, A. (1991). "Application of ion-selective electrodes in clinical analysis." *Electroanalysis.* 3: 727–34.

Sergeant, E.P. (1984). *Potentiometry and Potentiometric Titrations.* Wiley, New York. pp. 129–32.

Umezawa, Y. (1990). *Handbook of Ion Selective Electrodes: Selectivity Coefficients.* CRC Press Boca Raton, Fl.

Werner, E. Morf (1981). *The Principles of Ion-Selective Electrodes and Membrane Transport.* Elsevier Scientific and Akademiai Kiado, Budapest. pp. 60–61.

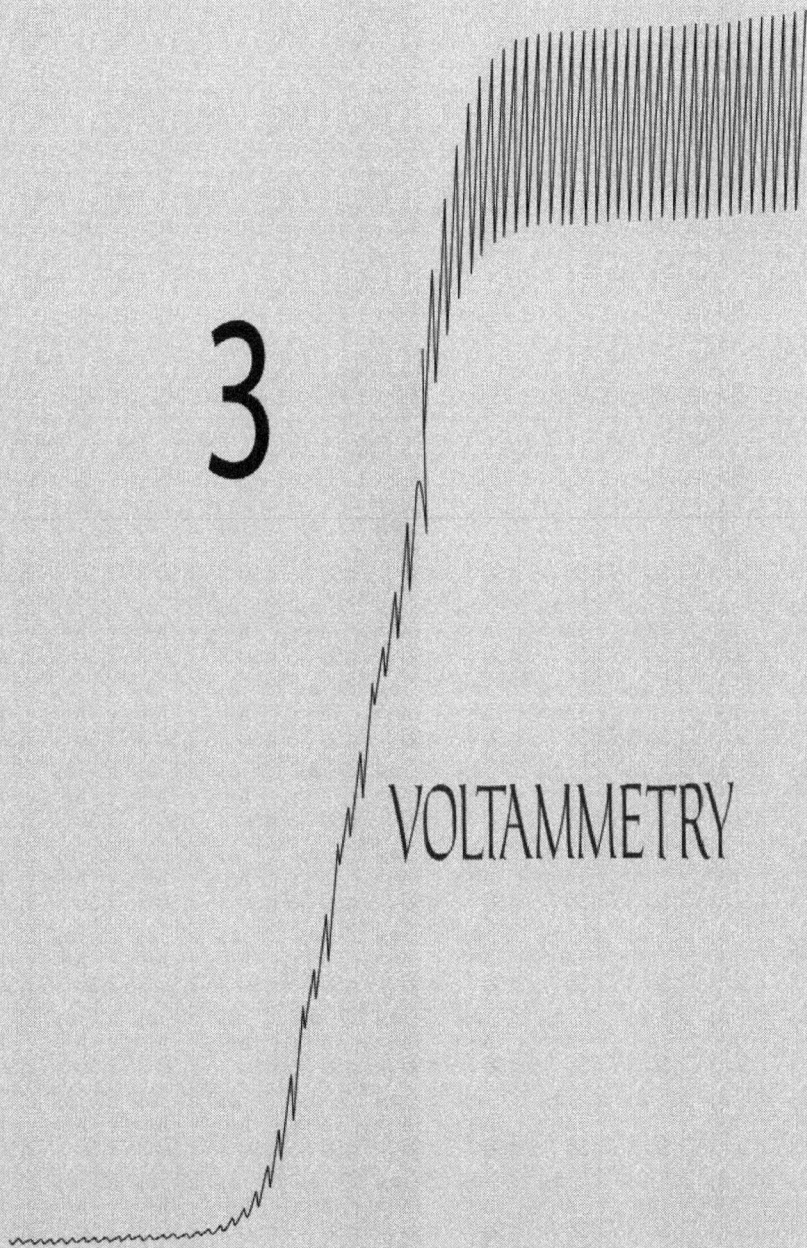

3

VOLTAMMETRY

Voltammetry is the study of potential–current–time relationship during electrolysis carried out in an electrochemical cell. The technique involves the measurement of current flowing in a cell by the applied potential, and in some cases, the variation of current with time is also studied.

The analytical advantages of the various voltammetric techniques include (i) excellent sensitivity with a wide linear concentration range for both organic and inorganic species (10^{-1} M to 10^{-12} M), for a large number of useful solvents and for electrolytes, (ii) simultaneous determination of several analytes, (iii) ability to determine various kinetic and mechanistic parameters, and (iv) rapid analysis time.

This technique is used as an analytical tool for the quantitative determination of both organic and inorganic species. This technique is also used for a variety of purposes which include fundamental studies of redox processes in various solvent media, surface absorption studies, electron transfer reaction mechanism, speciation and thermodynamic properties of solvated species. The voltammetric technique coupled with HPLC serves to be an effective tool for the analysis of complex mixtures in pharmaceutical industries.

The techniques which come under the general heading of voltammetry are

 i. polarography
 ii. stripping voltammetry
 iii. chronopotentiometry

In polarography and stripping voltammetry techniques, current is recorded under the applied potential, whereas in chronopotentiometry, variation of potential is measured with respect to time when constant current is applied.

POLAROGRAPHY

Conventional or direct current polarography refers to the use of the well-known current–voltage curves obtained by applying a DC potential between a dropping mercury electrode (DME) with gravity-controlled drop times of approximately 2–10 seconds and a reference electrode. Though the technique is used for the qualitative and quantitative determination of metals, it has no real rival for the determination of many heavy metals and of trace levels of organic compounds. Enormous advances in theory and instrumentation of polarography was developed for such determination. This modern and developed technique is now a sensitive, rapid technique applicable to analysis in the inorganic, organic, geochemical, biochemical, medical and pharmaceutical fields. Modern polarography technique includes phase-sensitive alternating current polarography, pulse polarography and linear sweep polarography.

DC POLAROGRAPHY
(CONVENTIONAL POLAROGRAPHY)

PRINCIPLE

A potential is applied to a cell consisting of a large quiescent mercury anode, a small mercury cathode (DME) and a dilute solution of electroactive material under examination. The metal ion in solution will not be deposited as a metal on the cathode, and no current flows until the potential of the cathode reaches the decomposition (discharge) potential of the electroactive species. As the potential increases above the discharge potential, the rate of reduction and the flow of current increases until a thin layer of solution around the cathode has been depleted of the ions being reduced. The current is controlled by the rate of diffusion of reducible ions through this layer. The rate of diffusion depends upon the concentration and diffusion coefficient of the ions in solution. The potential difference between the electrode–electrolyte interface is negligible and the rate of diffusion remains constant as the potential is increased. The limiting current known as diffusion current is proportional to the concentration of the ions in solution and hence the amount of ions present in the solution can be determined by comparing the diffusion currents in the solution and in a standard solution of known concentration. The **diffusion current** is measured from the height of the step in an *i–E* curve. The voltage at which half of the diffusion current is referred to as **half-wave potential** which is characteristic of a particular ion in solution. The current–potential curve is known as a polarogram as shown in Figure 3.1.

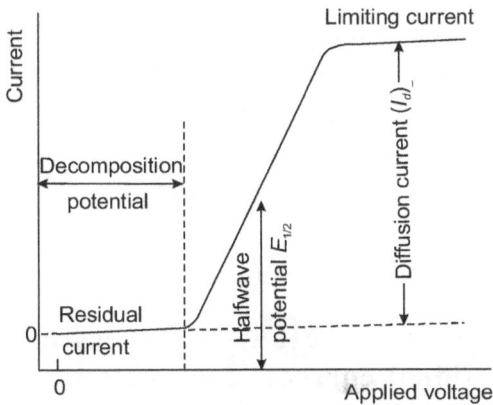

Figure 3.1 A typical polarogram

FARADAIC AND NON-FARADAIC PROCESSES

Two distinctly different types of processes, occur at electrodes. They are the faradaic and non-faradaic processes.

Faradaic Current

The process in which electrons are transferred across the electrode–solution interface and obeys Faraday's law, is called faradaic process. The magnitude of the faradaic current is governed by the electrode mechanism or mass transfer process and the rate of electrolysis is controlled by diffusion, electron transfer, chemical kinetics, adsorption, etc. These processes will be dealt in detail in the electrode mechanisms.

Non-Faradaic Current

Another type includes those in which charge transfer reactions (oxidation or reduction) do not occur across the electrode–solution interface because they are thermodynamically or kinetically unfavourable at all potential regions, whereas processes not involving electrolysis such as adsorption or desorption can occur. These processes are called non-faradaic processes because no electron transfer is involved. However, they can contribute current flow to the electrochemical cell. Thus, the total current flowing through the cell can be considered as the sum of faradaic and non-faradaic contributions.

POLAROGRAPHIC WAVE

The polarographic wave (Figure 3.1) consists of two regions namely, **residual current** (due to electrode polarization (charging current) or deposition of impurities) and **diffusion current** (due to mass transport by concentration gradient).

Charging Current

The current flowing between the electrodes when the applied potential is lower than the decomposition potential is called **residual current**. The residual current is caused by the discharge of solution impurites. At any potential, the magnitude of current indicates the energy required to set up the electrical double layer and to maintain this structure against the disordering effects of the growth of the drop and the Brownian motion in solution. The behaviour of this double layer remains analogous to that of a capacitor. The current needed to develop and maintain this double layer is known as **charging current** or **condenser current**.

In the case of stationary working electrodes, the double layer remains stable and the charging current remains same and neglible. But in the case of DME, when Hg drop falls, the double layer is removed and a fresh double layer is formed around the new drop. Each drop must therefore be charged to the required potential, thus a continuous flow of charging current makes the required potential appreciable. The magnitude of this current at DME is a few-tenth of μA in a deaerated solution of KCl.

Mass Transport

There are three modes of mass transport, viz., diffusion, migration and convection. The diffusion current is governed only by diffusion of species. This is achieved by minimizing current due to migration and convection.

Migration It is the movement of charged species due to a potential gradient and it is the mechanism by which charge passes through the electrolyte; the current of electrons through the external circuit must be balanced by the passage of ions through the solutions between the elctrodes (both cations to the cathode and anions to the anodes). It is however, not necessarily an important form of mass transport for the electroactive species even if it is charged. Hence, the migration current is assumed to be zero or negligible for an electroactive species when the polarogram is recorded in the presence of an indifferent electrolyte called as supporting electrolyte.

The value of the migration current depends on the transference number (t_j) of the species being reduced or oxidized.

$$t_j = \frac{c_j \lambda_j}{\sum\limits_i c_i \lambda_i}$$

where, c_j is the concentration of the electroactive species and λ_j is its equivalent conductance. The subscript i refers to any species present in the solution. The addition of supporting electrolyte, whose ions do not oxidize or reduce, causes the transference number of the electroactive species to decrease. If the concentration of the supporting electrolyte is 100 times higher than the electroactive species, the transference number, and hence the migration current of the species being reduced or oxidized becomes practically zero compared to the supporting electrolyte. Keeping the concentration of supporting electrolyte higher than the concentration of electroactive species also helps in decreasing the resistance of solution in the polarographic cell and in minimizing the iR drop across the electrodes.

Convection   It is the movement of a species due to mechanical forces. It can be eliminated by carrying out the electrolysis in a thermostat in the absence of stirring or vibration.

In industrial practice it is much more common to stir or agitate the electrolyte or to flow the electrolyte through the cell. These are all forms of forced convection and when present they have large influence on the current.

Diffusion It is the movement of the species under a concentration gradient, and it must occur whenever there is a chemical change at a surface. When an electrode reaction takes place (oxidation or reduction), the concentration of electroactive species becomes lower at the electrode surface than in the bulk. Hence, there is an existence of concentration difference; electroactive species diffuses from bulk to electrode surface layer.

Limiting Currents

Diffusion-controlled limiting currents It is assumed that all the ions or molecules that are initially present at the electrode surface will be electrolysed immediately since the potential is applied in the limiting current region. The concentration of electroactive species will approach zero in a very thin layer of solution near the electrode surface. Since the current flowing through the cell will be proportional to the quantity of material being electrolysed, the limiting current is proportional to the rate at which electroactive substances diffuse towards the electrode surface. Under this condition, the limiting current, i_1, is therefore referred to as a diffusion-controlled limiting current, i_d.

The diffusion current is determined by the concentration gradient at the surface of the electrode, i.e., $\left(\dfrac{\partial c}{\partial x} \right)_{x=0}$ which is time-dependent and is governed by Fick's law of diffusion.

Thus

$$i = nFAD\left(\frac{\partial c}{\partial x}\right)_{x=0}$$

where,

D is the diffusion coefficient of the species under examination,

A is the area of the electrode,

n is the number of electrons transferred,

F is the Faraday constant, and

$\left(\dfrac{\partial c}{\partial x}\right)_{x=0}$ is the concentration gradient.

By incorporating correction for stirring produced by the preceding drop fall off, the exact geometry of the electrode including shielding effects, etc. in the above equation, the diffusion current for a polarographic wave is given by the equation called Ilkovic equation.

$$i_d = 708n \; (C_0^* - C)D^{1/2}m^{2/3}t^{1/6}$$

where,

i_d is the diffusion-controlled current,

n is the number of electrons involved in charge transfer process,

C_0^* is the concentration of electroactive species in the bulk solution in mol/cm^3,

$D^{1/2}$ is the diffusion coefficient in cm^2/sec,

C is the concentration of electroactive species at electrode surface in mol/cm^3,

m is the flow rate of mercury in g/sec and

t is the drop time in sec.

The diagnostic criterion commonly employed in analytical work to ascertain that the limiting current is diffusion-controlled is the linear dependence on the square root of mercury column height. The mercury column height 'h' is the distance between the top of the mercury reservoir and the top of the DME.

Since t is directly proportional to $1/h$ and m is directly proportional to h.

$$i_d \propto m^{2/3}t^{1/6} \propto h^{1/2}$$
$$\therefore i_d \propto \sqrt{h}$$

Thus, a plot of i_d vs $\sqrt{h}$ should be linear, passing through the origin, if the limiting current is diffusion-controlled.

Reversible process   For a reversible electrode process, the shape of i–E curve at the potentials can be derived by combining the Nernst and Ilkovic equations. Consider a reversible process

$$A + ne^- \rightleftharpoons B$$

then the Nernst equation is given by

$$E = E^0 + \frac{RT}{nF} \ln \frac{[A]}{[B]}$$

The reactant A diffuses towards the electrode and the diffusion current is

$$i = 708 \, n([A]_0 - [A])D_A^{1/2} \, m^{2/3}t^{1/6}$$
$$i = i_d - 708 \, n[A]D_A^{1/2} \, m^{2/3}t^{1/6}$$
$$[A] = \frac{i_d - i}{708 \, nD_A^{1/2} \, m^{2/3}t^{1/6}}$$

After electrolysis, the product B diffuses towards the bulk solution and the diffusion current in this case is

$$i = 708n\{[B]_0 - [B]\}D_B^{1/2}\, m^{2/3}t^{1/6}$$

But $[B]_0 = 0$, hence

$$[B] = \frac{i}{708\, n\, D_B^{1/2}\, m^{2/3}t^{1/6}}$$

Substituting the concentrations of A and B in the Nernst equation

$$E = E^0 + \frac{RT}{nF}\ln\frac{i_d - i}{i}\frac{[D_B]^{1/2}}{[D_A]^{1/2}}$$

Since the diffusion coefficients of oxidized and reduced forms are equal, the equation can be written as

$$E = E^0 + \frac{RT}{nF}\ln\frac{i_d - i}{i}$$

when

$$i = \frac{i_d}{2},\quad E_{1/2} = E^0$$

$$E = E_{1/2} + \frac{RT}{nF}\ln\frac{id - i}{i}$$

This equation is called Heyrovsky–Ilkovic equation. A plot of E vs log $(i_d - i)/i$ should be linear with a slope $\dfrac{2.303RT}{nF}$ when $i = \dfrac{id}{2}$, log $(i_d - i)/i = 0$ and $E = E_{1/2}$. The $E_{1/2}$ of an electrode process can be calculated by the above method. The $E_{1/2}$ is a constant and independent of concentration of the electroactive species for a reversible electrode process.

Since the above method is a time-consuming procedure, an alternative approach is to simply measure the difference $E_{1/4} - E_{3/4}$ from the polarogram, where $E_{1/4}$ and $E_{3/4}$ correspond to value of E at $\frac{1}{4}i_d$ and $\frac{3}{4}i_d$ respectively.

$$E_{1/4} - E_{3/4} = \frac{59}{n} \text{ mV}$$

Irreversible process   For an irreversible process, the Nernst equation is no longer valid and the rate theory must be used. Consider an irreversible reaction,

$$A + ne^- \rightarrow B$$

and using mean currents, a solution is given by

$$E = E_{1/2} + \frac{RT}{\alpha nF} \ln \frac{i_d - i}{i}$$

Here, the current is no longer the diffusion-controlled value but is governed by the electron transfer rate. However, the limiting current is still the diffusion-controlled value.

α is independent of potential and the log plot is a straight line but the slope is larger than that in the reversible case because α lies between 0 and 1. Similarly values of $E_{1/4} - E_{3/4}$ are larger than those for the reversible case. $E_{1/2}$ is directly proportional to log t for a totally irreversible wave. The i vs $\sqrt{h}$ is a straight line only at the limiting plateau and not for any point on the wave.

The distinction between reversible and irreversible processes can be made therefore on the basis of wave shape, wave position and drop time dependence.

Kinetically controlled limiting currents   There are many cases in which the wave height is partially or wholly determined by the rate of a chemical reaction that produces an electroactive species near the electrode surface. Consider the following three mechanisms.

CE mechanism

$$Y \underset{k_1}{\overset{k_1}{\rightleftharpoons}} A$$

$$A + ne^- \rightleftharpoons B$$

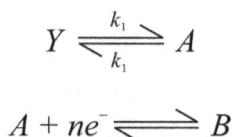

where k_1 and k_{-1} are the first-order or pseudo-first-order rate constants. The kinetically controlled electrode process depends on the ratio of the equilibrium concentrations of Y and A in the bulk solution and the values of k_1 and k_{-1}. If the concentration of A is low and k_1 is slow, the limiting current is governed by kinetically controlled limiting currents and if the equilibrium concentration of A is high, i the limiting current is governed by both diffusion and kinetically controlled limiting currents.

The limiting current is purely kinetically controlled, i.e., $i_k << i_d$

$$i_k = 0.493 \, n \, D^{1/2} [y] m^{2/3} t^{2/3} \frac{k_1}{k_{-1}^{1/2}}$$

Here $m \propto h$, $t \propto \dfrac{1}{h}$

Hence, $m^{2/3} t^{2/3}$ is independent of h.

Thus i_k is independent of mercury column height. This independency of i_k on mercury column height shows that the process is kinetically controlled.

A current–time plot of kinetically controlled wave obeys the following law.

$i = \text{constant} \times t^{2/3}$ at the limiting region.

This law holds good for current–time data obtained with the first drop as well as with the successive drops. Since the process is kinetically controlled, there is no diffusion limitation. Hence, concentration polarization is not transferred to the succesive drop.

EC mechanism

$$A + ne^{-} \rightleftharpoons B$$

$$B \underset{k_{-1}}{\overset{k_1}{\rightleftharpoons}} X$$

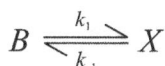

The EC mechanism is not influenced by the kinetics of the following chemical reaction. Hence, equations applicable to diffusion-controlled process remain valid for this case.

Catalytic mechanism

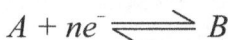

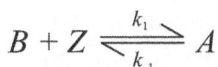

$$A + ne^{-} \rightleftharpoons B$$

$$B + Z \underset{k_{-1}}{\overset{k_1}{\rightleftharpoons}} A$$

Unlike CE mechanism where $i_l = i_k \ll i_d$, and EC mechanism where $i_l = i_d$, the catalytic current is always greater than that expected from the Ilkovic equation.

If the catalysing reagent z is present in excess in the solution, the rate constant of a pseudo-first-order is $k_1[z]$. If $k_1[z]$ is small, very little B will be reoxidized and the wave height will be equal to the diffusion current of A as measured in the absence of z.

If $k_1[z]$ is high, the chemical reaction produces more electroactive species near the electrode surface than that arising by diffusion from the bulk solution, so the $i_c >> i_d$ (where i_c is current due to chemical reaction and i_d is diffusion current.)

The dependence of catalytic currents on the mercury column height is complex. Two extreme cases are possible (i) when the sum of the forward and backward rate constants of the chemical reaction leading to the generation of the electroactive species is large, the limiting current is independent of column height (ii) when the above sum is very small, the limiting current is proportional to $\sqrt{h}$. Therefore the dependence of the limiting current on the mercury column height is not a useful criterion.

Adsorption controlled limiting currents   Electrodes can adsorb many organic compounds and this may influence the faradaic current. The following two processes occur whereby limiting current is influenced by adsorption.

   i. The electroactive species or their reduction product is adsorbed. A separate adsorption wave is formed. In addition to this wave, maxima, or minima or any other irregularity may occur in the limiting region.

  ii. Some other impurities in the solution are adsorbed. The presence of this substance at the electrode surface shift, deform or split the polarographic wave.

When DME is completely covered by absorption, the absorption current is given by

$$i_a = n\ Fz\, 0.85\ m^{2/3} t^{-1/3}$$

or

$$i_a = \text{constant} \times h$$

The dependence on *h* and independence on concentration are the two important criteria assigning adsorption currents.

POLAROGRAPHIC MAXIMA

Sometimes polarograms exhibit a peak. A maximum in the current could be due to the streaming of the electrolyte near the interface. If the streaming is as a result of differential polarization of the top and bottom of the mercury drop, then the phenomenon is referred to as polarographic maximum of the first kind. This appears as a sharp peak within a narrow potential range and is more pronounced at a lower concentration of the supporting electrolyte.

In instances where the streaming is induced by a fast flow of mercury in the capillary, it is known as polarographic maximum of the second kind. This appears over a longer potential and the peaks are rounded in appearance. These generally occur at higher concentration of electrolyte and are pronounced at high pressure of mercury. This type of maxima is otherwise called as non-streaming maxima.

The polarographic maxima of the first kind is eliminated by surfactants such as gelatine, methyl red or triton-X-100. About 0.002% of surfactant usually gives satisfactory results. The polarographic maxima of the second kind is readily distinguished by the fact that a decrease in the mass rate of flow in the capillary leads to an elimination of the maximum.

INSTRUMENTATION

The basic components of a modern electroanalytical system (Figure 3.2) for voltammetry are a potentiostat, computer and the electrochemical cell.

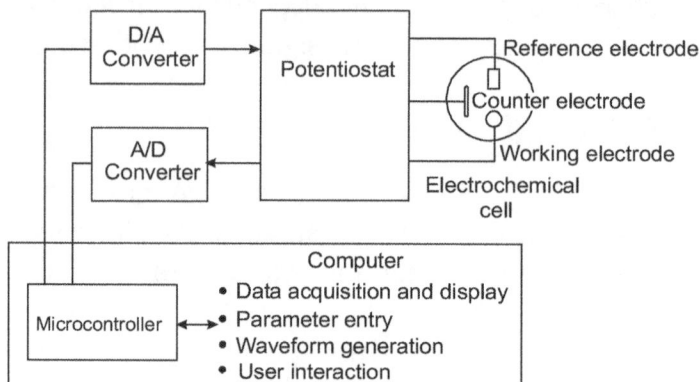

Figure 3.2   Block diagram of basic components of a voltammetric analysis

Potentiostat

The function of a potentiostat is to apply a known potential and to monitor the current. The most widely used potentiostats today are assembled from discrete integrated circuit, operational amplifiers and other digital modules. In many cases, especially in the larger instruments, the potentiostat package also includes electrometer circuits, A/D and D/A converters and dedicated microprocessors with memory.

A simple potentiostat circuit for a three-electrode cell with three operational amplifiers (OA) is depicted in Figure 3.3. The output of OA-1 is connected to the counter electrode with a feedback to its own inverting input through the reference electrode. This feedback decreases the difference between the inverting and non-inverting inputs of OA-1 and causes the reference electrode to assume the same potential as E_{in} of OA-1. Since the potential difference between the working electrode and the reference electrode is zero, the working electrode is set to the same potential as applied to the OA-1 input. With the reference electrode connected to E_{in} through the high impedence of OA-3, the current must flow through the counter electrode. Current flow through the reference is undesirable because of its high resistance, which would eventually cause its potential to become unreliable. A three-electrode system is normally used in voltammetry for currents in the range of μA to mA. With the use of microsized electrodes, currents are in the pA to nA range, and thus two electrodes are often used. An OA acting as a current-to-voltage converter (OA -2) provides the output signal for the A/D converter.

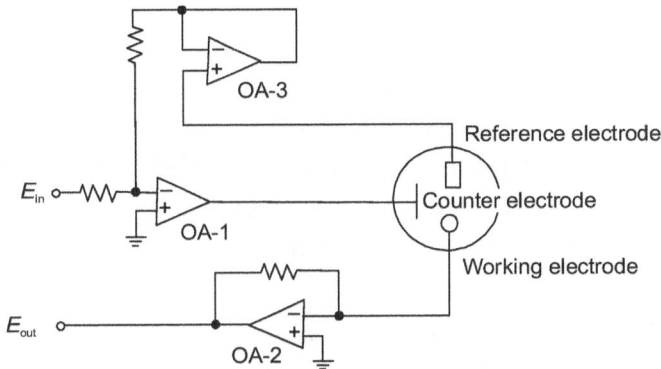

Figure 3.3 Potentiostatic circuit for three-electrode cell

Most voltammetric techniques are dynamic (i.e., they require a potential modulated according to some predefined waveform). Accurate and flexible control of the applied potential is a critical function of the potentiostat. In the early analog instruments, a linear scan just meant, a continuous linear change in the potential from one preset value to another. Since the advent of digital electronics, almost all the potentiostats have been operated in a digital fashion.

Thus the application of a linear scan is actually the application of a "staircase" modulated potential with small-enough steps to be equivalent to the analog case. Not surprisingly, digital fabrication of the applied potential has opened up a whole new area of pulsed voltammetry which speeds up the experiments and increases sensitivity. In the simplest standalone potentiostats, the excitation signal that is used to modulate the applied potential is usually provided by an externally adjustable waveform generator. In the computer-controlled instruments, the properties of the modulation and the waveform are under software control and can be specified by the operator. The most commonly used waveforms are linear scan, differential pulse, and triangular and square wave.

The use of micro- and nanometre-sized electrodes has made it necessary to build potentiostats with very low current capabilities. Microelectrodes routinely give current responses in the pico- to nanoampere range. High-speed scanning techniques such as square-wave voltammetry require very fast response times from the electronics.

ELECTROCHEMICAL CELL

A typical electrochemical cell consists of the sample dissolved in a solvent, an ionic electrolyte and three (or sometimes two) electrodes.

Two-electrode Polarograph

This consists of a working electrode (DME) and a large counter electrode. The potential is applied across the entire cell rather than across the working electrode–solution interface. Data recorded with this system are considerably in error if solution resistances and the resultant ohmic iR drop across them

(DME and counter electrode) are significant. Thus polarography in non-aqueous solvents is not possible with their two-electrode polarograph.

Three-electrode Polarograph

It consists of a working electrode (DME), reference electrode and a counter electrode with the external circuit arranged so that potential control is maintained between the DME and the reference electrode. Therefore the cell current passes between the DME and the auxiliary electrode (counter electrode). The potential of the reference electrode relative to the solution will not change with the current flow. The total potential applied to the cell will consist of the constant potential of the reference electrode, a potential between the solution and the working electrode, and a potential through the solution owing to its resistance. This ohmic resistance of the solution can be avoided by using the solution having either a low specific resistance or by placing the working and reference electrodes very close together.

The sample holders (cells) come in a variety of sizes, shapes and materials. The type used depends on the amount and type of sample, the technique and the analytical data to be obtained. The material of the cell (glass, teflon, polyethylene, etc.) is carefully selected to minimize reaction with the sample.

Reference electrodes The reference electrodes should provide a reversible half reaction with Nernst behaviour, be constant over time, and should be easy to assemble and maintain. The most commonly used reference electrodes are saturated calomel electrodes and Ag–AgCl electrode. They are available in a variety of sizes and shapes. (For more details, refer Chapter 2.)

Counter (auxiliary) electrodes   The analytical reactions
at the electrode surfaces occur at very short time periods and
rarely produce any appreciable change in bulk concentration
of reductant or oxidant. Thus, isolation of counter electrode
from the sample is not normally necessary. Thus Pt wire, Au
and graphite have also been used as counter electrodes.

Working electrodes   The working electrodes of various
geometries and materials range from small Hg drop to flat Pt
discs. Mercury is useful because it displays a wide negative
potential range, its surface is readily regenerated by producing

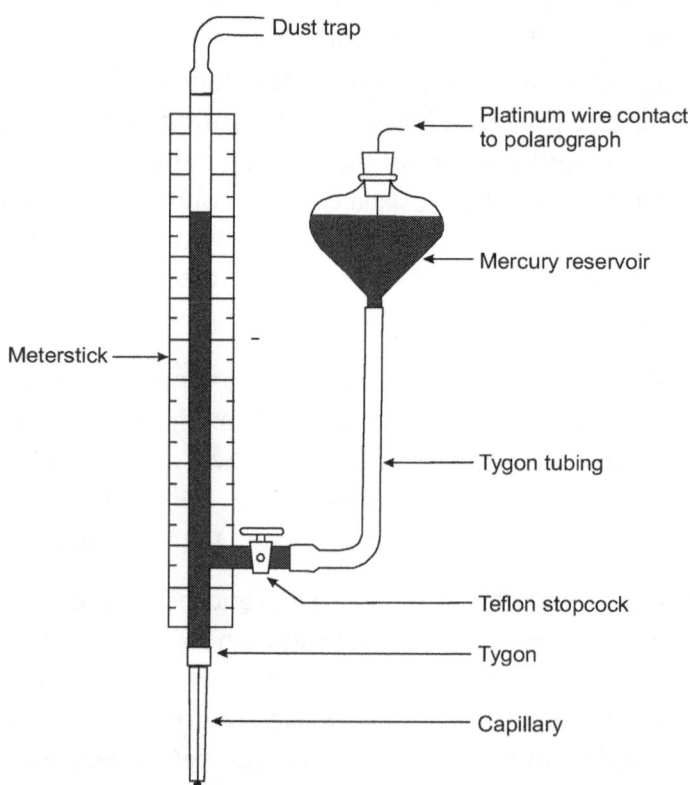

Figure 3.4   Dropping mercury electrode

a new drop or film and many metal ions can be reversibly reduced into it. Au, Pt and glassy carbon are also commonly used working electrodes. Even though polarography is considered as one of the techniques in voltammetry, it differs from other voltammetric methods both because of its unique place in the history of electrochemistry and due to its unique working electrode, the dropping mercury electrode (DME).

Dropping mercury electrode The arrangement of a dropping mercury electrode is shown in Figure 3.4. The DME consists of a glass capillary through which Hg flows under gravity from the Hg reservoir to form a succession of mercury drops. Each new drop provides a clean surface at which the redox process takes place, giving rise to a current that increases with increasing area as the drop grows, and then falling when the drop falls. The electrical contact is made with Hg by Pt wire.Figure 3.5 shows a polarogram for the reduction of Cd (II) in 1M $NaClO_4$. The drop time can be adjusted over a reasonably wide range by varying the height of the mercury reservoir. The drop time ranging from 2.9 seconds is used. A DME is characterized by its capillary constants such as (i) mercury flow rate (ii) drop time (t). The flow rate can be determined by allowing the Hg to flow through the capillary in a small weighed container. Once again the container is weighed and the mass of the Hg can be calculated. The measurements are made at the same height of the Hg column at which polarographic study is normally carried out. The drop time is generally measured in 0.1 M KCl solution at a fixed applied potential. It is then calculated from the time taken for 10–20 drops to fall from the capillary at the same height of the Hg column at which *m* is calculated. The drop time depends on the composition of the solution and also on the applied potential.

Even though many electrodes are available, the DME is the best for obtaining current–voltage curve because of the following advantages.

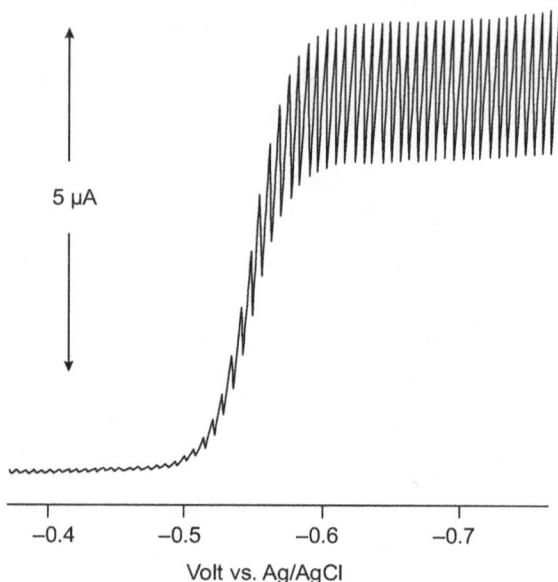

Figure 3.5   DC polarogram of reduction of Cd(II) in 1 M NaClO$_4$

i. It has a smooth and continually renewable surface exposed to the solution being analysed.

ii. Each drop is unaffected by the reactions which occurred at the surface of the preceding drops.

iii. It readily forms amalgam with almost all metals and then lowers their reduction potential.

iv. The diffusion equilibrium at the mercury–solution interface is rapidly attained.

v. The high hydrogen overvoltage of Hg enables analyses to be carried out in acid solution.

The disadvantages of DME are also listed here.

i. Mercury is toxic and should be handled carefully.

ii. Hg has limited applications in the more positive potential, since anodic dissolution of Hg starts at about 0.4 V.

iii. Change in applied potential causes a change in the surface tension of Hg and therefore changes the size of the drop.

iv. The addition of surface active agents changes the size of the drop.

Inert Atmosphere

At room temperature, solution of electroactive species contains 10^{-3} M dissolved oxygen. The dissolved oxygen gives two polarographic reduction waves over a wide range of working potential which interferes with most of the substances to be analysed. Oxygen can be removed by bubbling an inert gas like Ne or Ar through the solution.

QUANTITATIVE TECHNIQUES

The following three methods are widely used in analytical methods.

Wave Height–Concentration Plots

Solutions of several different concentrations of the ion under investigation are prepared, the composition of the supporting electrolyte and the amount of maximum suppressor added being the same for the comparison standards and for the unknown.

The heights of the waves obtained are measured in any convenient manner and plotted as a function of the concentration. The polarogram of the unknown is produced exactly as the standards and the concentration of the unknown can be taken from the graph. The method is strictly empirical, and no assumptions, except correspondence with the conditions of calibration, are made. The wave height need not be a linear function of the concentration, although this is frequently the case. For high-precision results, the unknown solution and standard solution should be recorded consecutively.

Internal Standard (Pilot Ion) Method

The relative diffusion currents of ions in the same supporting electrolyte are independent of the characteristics of the capillary electrode and, to a close approximation, of the temperature. Hence, upon determining the relative wave heights with the unkown ion and with some standard or 'pilot' ion added to the solution in known amounts, and comparing these with the ratio for known amounts of the same two ions, previously determined, the concentration of the unknown ion may be established. This procedure has a limited application, primarily because only a small number of ions are available to act as pilot or reference ions. The main requirement for such an ion (if singly charged) is that its half-wave potential should differ by at least 0.4 V from the unknown or any other ion in the solution with which it might interfere. When a single unknown is present, this condition can usually be satisfied, but in complex mixtures there is sufficient difference between the half-wave potentials to introduce additional waves.

Method of Standard Addition

The polarogram of the unknown solution is first recorded, after which a known volume of a standard solution of the same ion is added to the cell and a second polarogram is recorded. From the magnitude of the heights of the two waves, the known concentration of ion added, and the volume of the solution after the addition, the concentration of the unknown may readily be calculated. If I_x is the observed diffusion current of concentration C_x, and I_s is the observed diffusion current of the unknown solution of volume V ml and of current after V ml of a standard solution of concentration C_s has been added, then according to Ilkovic equation,

$$I_x = kC_x$$

and

$$I_s = k(VC_x + vC_s)(V + v)$$

Thus

$$k = I_s(V + v) / (VC_x + vC_s)$$

Hence,

$$C_u = \frac{I_x vC_s}{(I_3 - I_x)(V + v) + I_x v}$$

The accuracy of the method depends upon the precision with which the two volumes of solution and the corresponding diffusion currents are measured. The material added should be maintained in a medium of the same composition as the supporting electrolyte, so that the latter is not altered by the addition.

PULSE POLAROGRAPHY

BASIC PRINCIPLE

In the DC polarography, a potential is applied to the cell and the resultant current is measured. In the case of pulse polarography, the potential is applied periodically during short time intervals. Both the format for application of the pulse and the current read-out are varied in different ways and accordingly, the techniques are termed as **normal**, **derivative** and **differential** pulse polarography.

In order to understand how pulse polarography is useful in minimizing the measurement of capacitive current, let us consider an electrode maintained at a potential at which no faradaic reaction occurs. The current flowing at DME will be that caused by an increase in the double layer capacitance. At the end of the drop life when the rate of drop-growth is minimal, this residual current will be small and changes only very slowly with time. Thus, prior to the application of the pulse, a small but finite charging current attributed to the DC term exists in pulse polarography.

If a pulse is applied to the electrode, the potential is suddenly increased to a new value (there is still no faradaic reaction occurring) and then current must flow to charge the double layer to the new potential. Assuming the model of an ideal capacitor, the pulse-charging current will be the largest, immediately after application of the pulse and then will decay exponentially with time. When the amplitude of the pulse is sufficiently large, a reduction of electroactive species occurs and hence a faradaic current flows. If the potential of the pulse corresponds to a point on the raising part of the DC polarogram, the magnitude of the current will depend on the charge-transfer kinetics or rate-determining steps of any electrode process. When pulse is applied, there will be a large current jump and it decays as a function of time. This decay is much slower than the decay of charging current.

In this technique, only a single pulse is applied to the system per mercury drop, late in the drop life. It is assumed that the changing current has decayed to almost zero value of the 20 to 40 millisecond of application of the pulse. The current remaining (faradaic) after this time is measured. The polarogram is a plot of faradaic current produced by the pulse versus applied potential.

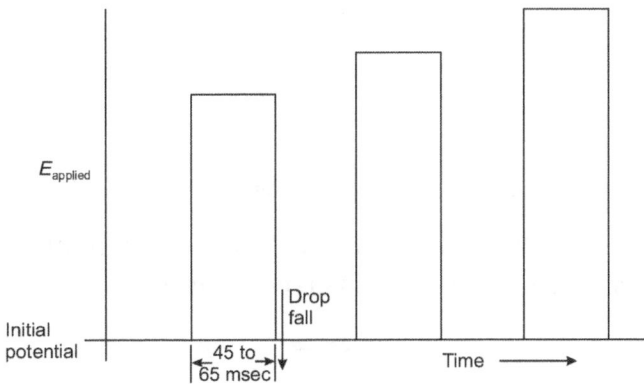

Figure 3.6   Applied potential waveform in normal pulse polarography

NORMAL PULSE POLAROGRAPHY

The potential pulses (E_{app}) of gradually increasing amplitude are applied to an electrode, starting from an initial potential where no faradaic current flows. The potential pulses are ~ 45 to 65 msec. duration applied to late in the drop life and the potential between pulses returns to the initial value. The applied potential waveform in normal pulse polarography is shown in Figure 3.6. The normal pulse polarograph is shown in Figure 3.7.

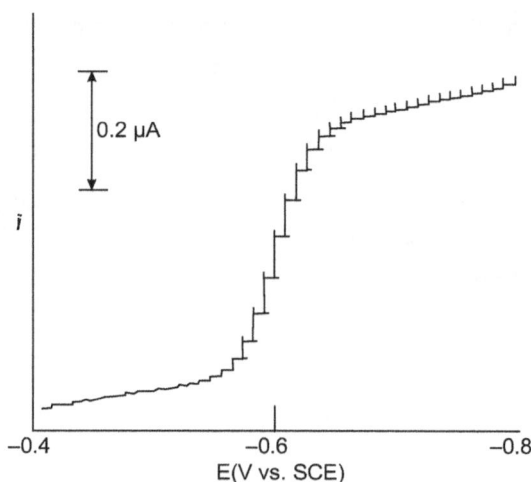

Figure 3.7   Normal pulse polarogram of Cu(II) in I M NaNO$_3$

Reversible and irreversible processes   The limiting current for any reversible process is given by the Cottrell equation.

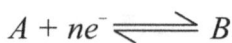

$$A + ne^- \rightleftharpoons B$$

$$i_l = nFCA\sqrt{\frac{D}{\pi t_m}}$$

where,

n is the number of electrons transferred

F is the Faraday constant

C is the concentration of the electroactive species

A is the area of the electrode

D is the diffusion coefficient of the electroactive species

t_m is the time interval between the pulse application and current measurement

The equation to the *i–E* curve for the reversible process may be written as

$$E = E_{1/2}^r + 2.303 \frac{RT}{nF} \log \frac{i_l - i}{i}$$

The plot of *E* vs $\log \frac{i_l - i}{i}$ is linear with slope $2.303 \frac{RT}{nF}$.

From the above equation, we understand that the diagnostic criteria for reversibility under normal pulse polarography is similar to that of DC polarography.

For a totally irreversible process, the equation to the *i – E* curve is given by

$$E_{1/2} = E_{1/2}^r + \frac{2.303RT}{\propto nF} \log 2.31 k_s \sqrt{\frac{t_m}{D}}$$

$E_{1/2}$ is a function of t_m for an irreversible process.

The Cottrell equation is applied equally well to reversible and irreversible processes and for a reduction process, and is given by

$$(i_l)_{red} = nFAC \sqrt{\frac{D}{\pi t_m}}$$

However, the limiting current for oxidation process $(i_l)_{ox}$ is obtained by commencing scanning from the potential where the reduction process diffusion plateau appears and proceeds to more positive potentials. The re-oxidation of the reduction product occurs during this scan. The current measured represents a difference in the current measured before and after the application of pulse.

Therefore, the current corresponding to positive scan will yield a limiting current corresponding to the DC current flowing prior to pulse application. Thus

$$(i_l)_{ox} = -nFAC\sqrt{\frac{7D}{3\pi t}}$$

The ratio of the limiting current for the normal and reverse scans for an irreversible pulse is given by

$$\frac{(i_l)_{red}}{(i_l)_{ox}} = \sqrt{\frac{3t}{7t_m}}$$

If t is greater than t_m by a factor of 100, the preceding ratio will have a magnitude of 7.

For a totally reversible process, it can be given as

$$\frac{(i_l)_{red}}{(i_l)_{ox}} = \left[1 - \sqrt{\frac{7t_m}{3t + 7t_m}} + \sqrt{\frac{7t_m}{3t}}\right]^{-1}$$

$$\text{since } t >> t_m$$

$$\frac{(i_l)_{red}}{(i_l)_{ox}} = 1$$

The wave height ratio of 7 : 1 for irreversible and 1 : 1 ratio for the reversible process provides an indication for the nature of the electrode process. In the case of reversible processes, the $E_{1/2}$ values for both forward and reverse scans are the same, i.e.,

$$\left(E_{1/2}\right)_{\text{red}} - \left(E_{1/2}\right)_{\text{ox}} = \frac{RT}{\alpha nF}\left[0.574 + \frac{1.49\ tm}{t}\right]$$

and the $E_{1/2}$ value for the reverse scan is several millivolts more negative than that for the forward scan. The scan reversal pulse polarography for both reversible and irreversible processes are shown in Figures 3.8 and 3.9.

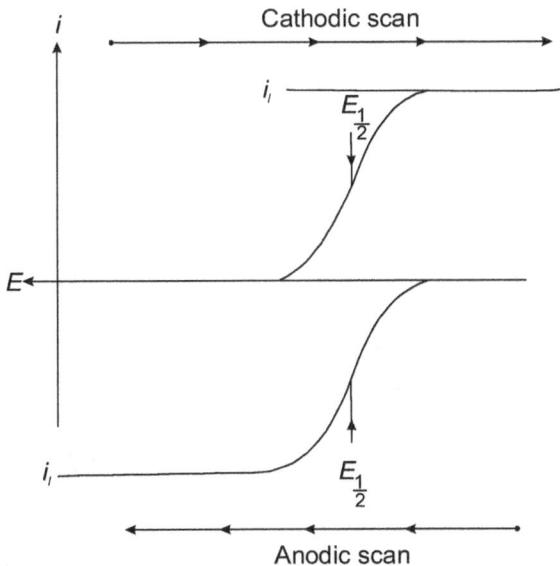

Figure 3.8   Scan reversal pulse polarography for a reversible reduction

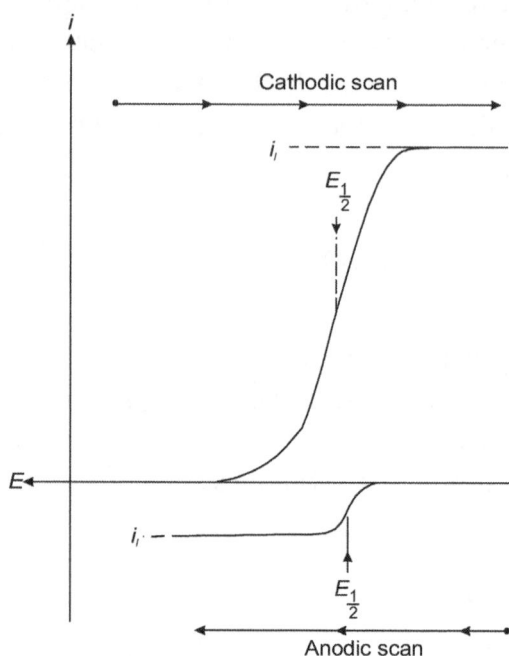

Figure 3.9   Scan reversal pulse polarography for an irrversible reduction

DIFFERENTIAL PULSE POLAROGRAPHY

In this technique, a normal DC voltage ramp (when applied DC voltage increases with time, it is called ramp) is applied to the system. Near the end of the drop life, a small amplitude pulse of approximately 50 mV is superimposed on to the ramp.

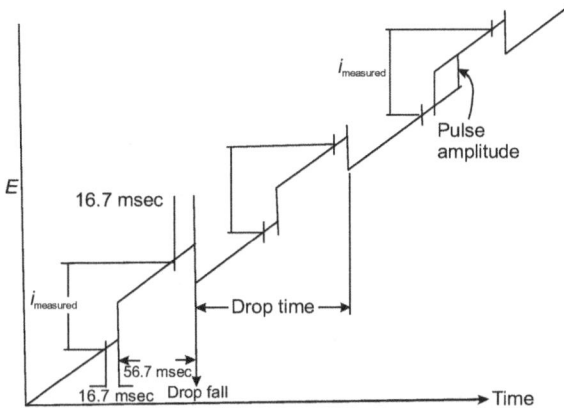

Figure 3.10 Applied potential waveform in differential pulse polarography

As the measured signal is the difference in the current measured before and after the application of the pulse, i.e., the change in the current produced by the perturbation of the system, a peak-shaped curve is obtained with the peak maximum occurring near $E_{1/2}$ if the perturbation is sufficiently small. The differential pulse polarogram is shown in Figure 3.11.

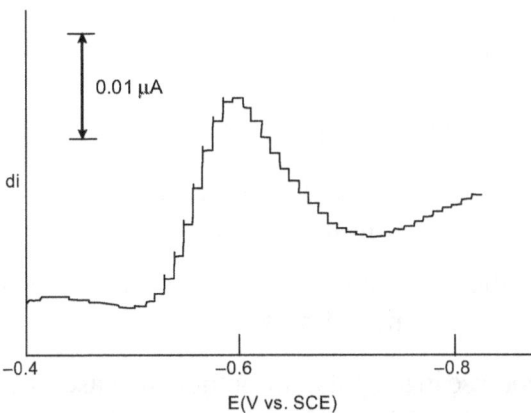

Figure 3.11 Differential pulse polarogram of Cu(II) in 1 M NaNO$_3$

Reversible and irreversible processes　The limiting diffusion current for a reversible process is given by Cottrell equation as

$$\Delta i = \frac{n^2 F^2}{RT} AC(-\Delta E) \sqrt{\frac{D}{\pi t_m}} \frac{p}{(1+p)^2}$$

where,　Δi = differential pulse current

ΔE = pulse amplitude

This equation is valid only in the case of small amplitudes. The following solution is valid for all values of ΔE.

$$\Delta i = nFAC \sqrt{\frac{D}{\pi t_m}} \frac{P_A \sigma^2 - P_A}{(\sigma + P_A \sigma^2 + PA + PA^2 \sigma)}$$

where

$$P_A = \exp \frac{nF}{RT} \left[\frac{E_1 + E_2}{2} - E_{1/2} \right]$$

$$\sigma = \exp \frac{nF}{RT} \left[\frac{E_2 - E_1}{2} \right]$$

$E_2 - E_1 = \Delta E$, the pulse amplitude.

E_2 = the potential at which the current i_2 is measured after the application of pulse.

E_1 = the potential at which the current i_1 is measured in the absence of pulse.

An increase in the pulse amplitude increases the peak width. The peak half-width is defined as the width of the peak at the point where the peak current is half its maximum height.

The diagnostic criteria for various electrode processes are given below.

Reversible process   Peak width, $w_{1/2}$ is $\dfrac{90.4}{n}$ mV at 25°C.

Quasi-reversible and totally irreversible process $(\Delta t)_{max}$ is a function of k_s and current per unit concentration is lower than the reversible process.

Catalytic and other electrode processes i vs C plots are curved as in DC polarography.

PULSE VOLTAMMETRY AT STATIONARY ELECTRODE

Normal pulse voltammetry is employed to study the electrode processes at stationary electrodes. At the beginning of a scan, the intitial potential chosen is inadequate to cause the reduction of the A and hence B is produced.

$$A + ne \rightleftharpoons B$$

at the electrode surface. Between the pulses, the potential returns to the initial value where reduction does not occur. If a process is reversible, reoxidation of B to A will occur at the rest potential and renew the surface. If the reduction product is a solid or an adsorbed species, renewal of electrode surface does not occur. Stirring of a solution can eliminate the depletion effect for reduction. Hence, pulse voltammetric analyses of irreversible systems should be carried out in the presence of stirring.

The diagnostic criteria for various electrode processes at stationary electrode are analogous to that for the DME.

APPLICATIONS

An antibiotic, chloramphenicol, of 1.3×10^{-5} concentration is the lower limit for the DC method. The wave is clearly discernible. The accurate evaluation of $E_{1/2}$ and wave height is highly impossible. But differential pulse voltammetry shows a sharp peak and allows precise measurement of the peak potential and the peak height. Quantitative and qualitative analysis of trace residues of any material is obtained because of the high sensitivity, and this technique has already been used in analytical methods.

FUNDAMENTAL HARMONIC AC POLAROGRAPHY

In sinusoidal AC polarography, a small amplitude sinusoidal alternating potential is superimposed on to the usual potential ramp used in polarography. After filtering at the DC component of the experiment, a plot of alternating current vs direct potential is made and this constitutes an AC polarogram. A conventional AC polarogram involves application of a potential containing a small sinusoidal component of fixed frequency and amplitude (ΔE) usually in the range 10 to 50 mV together with a DC component E_{dc}. The potential E of the cell is therefore given by the sum of AC and DC components as given by

$$E = E_{dc} - \Delta E \sin \omega t$$

The experimental conditions are the same as those used in conventional DC polarography and the supporting electrolyte employed should have a concentration 100 times higher than that of the electroactive species of interest. These conditions enable the theory to be solved in terms of the usual rate-determining steps such as diffusion control, heterogeneous electron transfer, homogeneous chemical reactions coupled to the electron transfer step or charging of the electrode double layer, etc.

The alternating current whose frequency is the same as that of the applied alternating potential is measured as a function of DC potential. The resulting plot of the fundamental harmonic alternating current $[I(\omega t)]$ vs DC potential constitutes the conventional DC polarogram. As with all polarographic methods, $I(\omega t)$ contains both faradaic and charging current components. The faradaic current provides a peak-shaped curve which coincides with the rising part of the DC polarographic wave for a reversible process. A typical AC polarogram of Cd(II) in Na_2SO_4 is shown in Figure 3.12. The peak current I_p of the AC polarogram is a linear function of concentration and is normally used as the basic parameter in quantitative analytical applications. The peak potential E_p corresponding to the value of I_p is closely related to $E_{1/2}$ and is characteristic of the electroactive species and the medium. The magnitude, shape and position of the faradaic wave are governed by the kinetic and thermodynamic properties of the electrode process.

The higher order current components are found at frequencies which are integral multiples of the fundamental frequency as well as at zero frequency. Thus by developing an appropriate system of measurement, second, third, fourth and higher order harmonic AC polarograms can be measured. In this chapter fundamental AC polarogram alone is discussed.

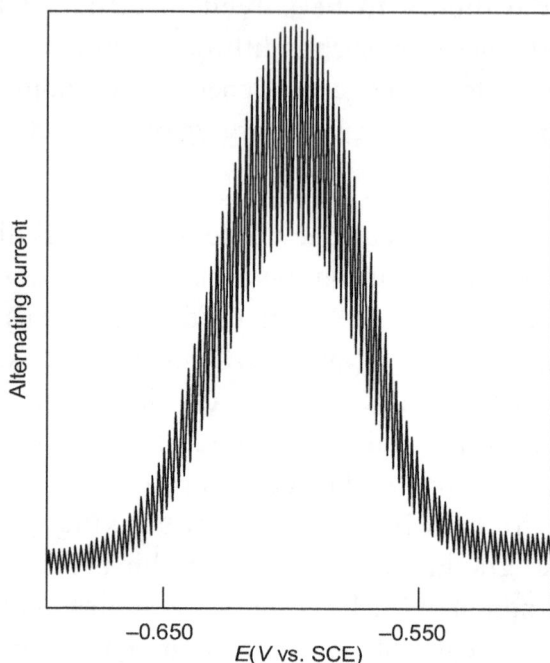

Figure 3.12   AC polarogram of Cd(II) in I M Na_2SO_4

FARADAIC ELECTRODE PROCESSES

The faradaic electrode processes can be considered in the usual four classes.

1. Reversible AC electrode processes
2. Quasi-reversible AC electrode processes
3. Irreversible DC electrode processes
4. AC electrode processes with coupled chemical reactions or adsorption

Reversible AC electrode processes   The current produced by a fundamental harmonic reversible wave $I(\omega t)$ is given by

$$I(\omega t) = \frac{n^2 F^2 AC(\omega D)^{1/2} \Delta E}{4RT \ \cos h\left(\dfrac{j}{2}\right)} \sin\left(\omega t + \frac{\pi}{4}\right)$$

where,

A = area of the electrode

C = concentration of electroactive species

ω = angular frequency

D = diffusion coefficient of electroactive species

ΔE = amplitude of applied alternating potential

t = time

j = $\dfrac{nF}{RT}\left(E_{dc} - E_{1/2}\right)$

E_{dc} = DC component of potential

$E_{1/2}$ = reversible half-wave potential

The i–E relation of the AC reversible wave is given as

$$E_{dc} = E_{1/2} + \frac{2RT}{nF}\ln\left[\left(\frac{I_p}{I}\right)^{1/2} \pm \left(\frac{I_p - I}{I}\right)^{1/2}\right]$$

It can be shown that the peak potential of wave, I_p, is equal to $E_{1/2}$. E_{dc} at half-wave height where $I = \dfrac{I_p}{2}$ is given by

$$E_{dc} = E_{1/2} + \frac{2RT}{nF}\ln\left(\sqrt{2} - 1\right)$$

or

$$E_{dc} = E_{1/2} + \frac{2RT}{nF}\ln\left(\sqrt{2} + 1\right)$$

where, the two solutions correspond to the two equivalent parts of the symmetrical wave. Subtraction of these two equations gives the width of the AC wave at half its height.

$$\text{Half-width} = \frac{2RT}{nF}\ln\frac{\sqrt{2}+1}{\sqrt{2}-1} = \frac{4RT}{nF}\ln\sqrt{2}+1$$

$$= 1.52\left(2.303\frac{RT}{nF}\right)$$

At 25°C, the above equation has a value close to $90/n$ mV. The measurement of half-width provides a criteria for the reversibility of fundamental harmonic AC electrode process. A graphic plot of E_{dc} vs $\log\left[\left(\frac{I_p}{I}\right)^{1/2} \pm \left(\frac{I_p - I}{I}\right)^{1/2}\right]$ should

be a straight line of slope $2\left(2.303\frac{RT}{nF}\right)$ for a reversible process. The following are the criteria for reversible process.

   i. The peak potential is independent of concentration and drop time.

   ii. I_p is linearly dependent on area, concentration, and ΔE.

   iii. I_p is independent of mercury column height.

Quasi-reversible AC electrode processes   The quasi-reversible wave will be slightly broader than that of the reversible electrode process with slight departure from the theoretical Nernstian slope $2.303\frac{RT}{nF}$ of the DC plot of E_{dc} vs $\log\frac{(i_d - i)}{i}$. The E_p varies with frequency and approaches $E_{1/2}$ at low frequency and for $k_s \geq 10^{-2}$ cm/sec.

Irreversible AC electrode process A measurable AC wave of magnitude and shape, which is independent of k_s will be observed with irreversible systems. The influence of k_s is to determine the position of the wave on the DC potential axis. The current magnitude $I(\omega t)$ is proportional to α, the charge transfer coefficient. The E_p value of an irreversible process is displaced in the negative direction from the DC-half-wave potential. This criterion characterizes an irreversible AC reduction wave. The waves are extremely broad and are of very low sensitivity (i.e., low current per unit concentration) compared with those of reversible AC electrode processes.

Adsorption process Adsorption and other surface phenomena, when coupled with charge transfer, can lead to highly distorted waves. Such waves are usually not analytically useful.

CHRONOPOTENTIOMETRY

A constant current is applied between working and auxiliary electrodes with the help of a galvanostat. The resulting change in potential is recorded as a function of time. The plot of potential vs time is a chronopotentiogram shown in Figure 3.13 which finds variety of analytical applications or investigation of electrode kinetics. The chronopotentiogram of a simple reversible electrode process $A + ne \rightleftharpoons B$ is shown in Figure 3.13. Before the application of constant current, the solution contains only A and it has relatively positive (oxidizing) potential. When current is applied, electrolysis begins, reactant A is reduced to B and working electrode becomes more negative (reducing) potential. If the electrode process is reversible, the working electrode obeys Nernst equation.

$$E = E^0 + \frac{RT}{nF} \ln \frac{C_A}{C_B}$$

The potential changes only slowly with time during the period that the ratio $\frac{C_A}{C_B}$ reaches 1. As the electrolysis continues, the concentration of A decreases to the point where there is insufficient A to accommodate all of the constant current, and A diffuses towards the electrode surface (unstirred solution and sufficient supporting electrolyte present to prevent migration, hence only diffusion current). The concentration of A near the electrode surface drops towards a negligible value, and the current forces the electrode to a potential at which a different reaction can occur. At this stage, a rapid change in potential is observed. The time from the start of the electrolysis until the rapid potential change is designated as transition time, τ.

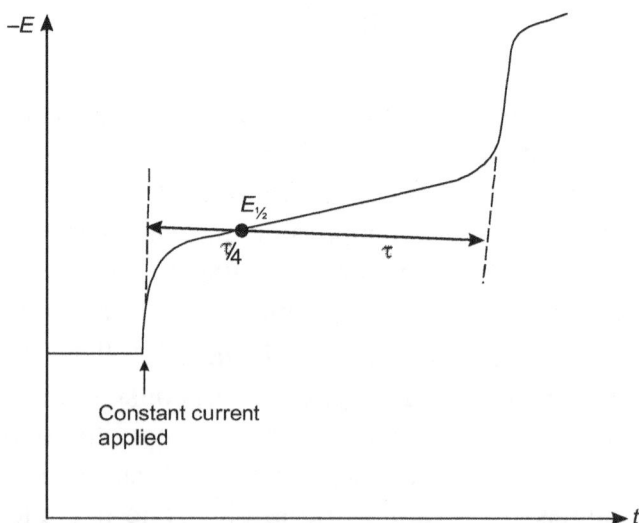

Figure 3.13   Chronopotentiogram for a simple electrode process

Under conditions of linear diffusion in an unstirred solution, the transition time (τ) is given by Sand's equation.

$$\tau^{1/2} = \frac{\pi^{1/2}nFAD^{1/2}C_A}{2i}$$

where,

n is the number of electrons transferred

F is the Faraday constant in coulombs

A is the area of the electrode

D is the diffusion coefficient of the reactant

C_A is the concentration of the electroactive species and

i is the constant current

$\tau^{1/2}$ is proportional to concentration and independent of electrode kinetics (k_s). The shape of the *E–t* curve in the controlled-current method is dependent on the electrode kinetics, as the *i – E* curve in polarography.

Consider a simple reversible electrode process occurring at a planar electrode under linear diffusion condition. Initially *A* is present at concentration C_A.

$$A + ne^- \rightleftharpoons B$$

the equations governing the system are

$$\frac{\partial C_A}{\partial t} = D_A \frac{\partial^2 C_A}{\partial x^2}$$

$$\frac{\partial C_B}{\partial t} = D_B \frac{\partial^2 C_B}{\partial x^2}$$

At $t = 0$; $C_A = C_A$ and $C_B = 0$

At $x \to \infty$; $C_A \to C_A$ and $C_B = 0$

The flux condition is

$$D_A \left[\frac{\partial C_A}{\partial x} \right]_{x=0} + D_B \left[\frac{\partial C_B}{\partial x} \right]_{x=0} = 0$$

and the concentration gradient condition is

$$D_A \left[\frac{\partial C_A}{\partial x} \right]_{x=0} = \frac{i(t)}{nFA}$$

A concentration gradient condition replaces the concentration condition of the controlled potential methods.

If $i(t) = i$ (constant), then

$$C_A(0,t) = C_A - \frac{2it^{1/2}}{nFAD_A^{1/2}\pi^{1/2}}$$

and

$$C_B(0,t) = \frac{2it^{1/2}}{nFAD_B^{1/2}\pi^{1/2}}$$

Substituting the concentrations of A and B in Nernst equation, for a reversible process, the potential is given by

$$E = E^0 + \frac{RT}{nF} \ln \frac{C_A - t^{1/2} \left(\dfrac{2i}{nFAD_A^{1/2}\pi^{1/2}} \right)}{t^{1/2} \left(\dfrac{2i}{nFAD_B^{1/2}\pi^{1/2}} \right)}$$

When the electrolysis is complete, $C_A = 0$ and $t = \tau$, substituting this condition into expression for $C_A(0, t)$ gives the

Sand equation. Using the Sand equation to obtain the expression for C_A gives

$$E = E^o + \frac{RT}{nF} \ln \frac{\tau^{1/2} - t^{1/2}}{t^{1/2}} \left(\frac{D_B}{D_A} \right)^{1/2}$$

it can be written as

$$E = E_{\tau/4} + \frac{RT}{nF} \ln \frac{\tau^{1/2} - t^{1/2}}{t^{1/2}}$$

where, $E_{\tau/4} = E^o + \frac{RT}{nF} \ln \left(\frac{D_B}{D_A} \right)^{1/2}$.

When $t = \frac{\tau}{4}$, $E =$ reversible polarographic half-wave potential $E_{1/2}$.

The diagnostic criteria for a reversible process is a plot of E vs $\log \left[z^{1/2} - t^{1/2} / t^{1/2} \right]$ is a straight line with a slope of $2.303 \frac{RT}{nF}$ or $E_{1/4} - E_{3/4} = \frac{47.9}{n}$ mV at 25°C.

For a totally irreversible electrode process

$$E = E^o + \frac{RT}{\alpha nF} \ln \frac{nFAC_A k_s}{i} + \frac{RT}{\alpha nF} \ln \left(\frac{\tau^{1/2} - t^{1/2}}{\tau^{1/2}} \right)$$

The diagnostic criteria for an irreversible process is a linear plot of E vs $\log \left[1 - \left(\frac{t}{E} \right)^{1/2} \right]$ with a slope of $\frac{2.303RT}{\alpha \, nF}$ and $\left| E_{1/4} - E_{3/4} \right| = \frac{33.6}{\propto n}$ mV.

If the reversible process is followed by a chemical reaction, the entire $E - t$ curve is displaced to more positive potentials.

LINEAR SWEEP VOLTAMMETRY AND RELATED TECHNIQUES

This technique has become an important and widely used electroanalytical technique in many areas of chemistry. It is rarely used for quantitative determinations, but it is widely used for the study of redox processes, for understanding the reaction intermediates, and for obtaining the stability of reaction products.

This technique is based on utilizing a rapid linear sweep of the potential. When potential is applied to the dropping mercury electrode (DME), this method enables the entire potential range to be covered on one drop, i.e., the voltage sweep time is short compared with the drop time. In polarography, it is assumed that constant potential–current curves are being recorded because the voltage sweep time is much greater than the drop time, and the potential covered per drop is only in the order of mV or less. In the linear sweep method the theory is solved under conditions of a continuously changing potential.

The recording of the $i - E$ curve at fast scan rates, requires the use of an oscilloscope, and so the technique of linear sweep voltammetry (LSV) has been referred to as **oscillographic polarography**. It can also be referred to as **stationary electrode polarography**. The method LSV at DME is the one in which the scan rate occurs over only a fraction of the life of the mercury drop. The potential is considered to have been applied to a stationary electrode because during the recording of the $i - E$ curve, the growth of the drop is negligible. The theory discussed for LSV at a DME, therefore, can be applied to any stationary electrode, e.g. Pt, carbon, gold, etc. are used when mercury electrodes are not suitable.

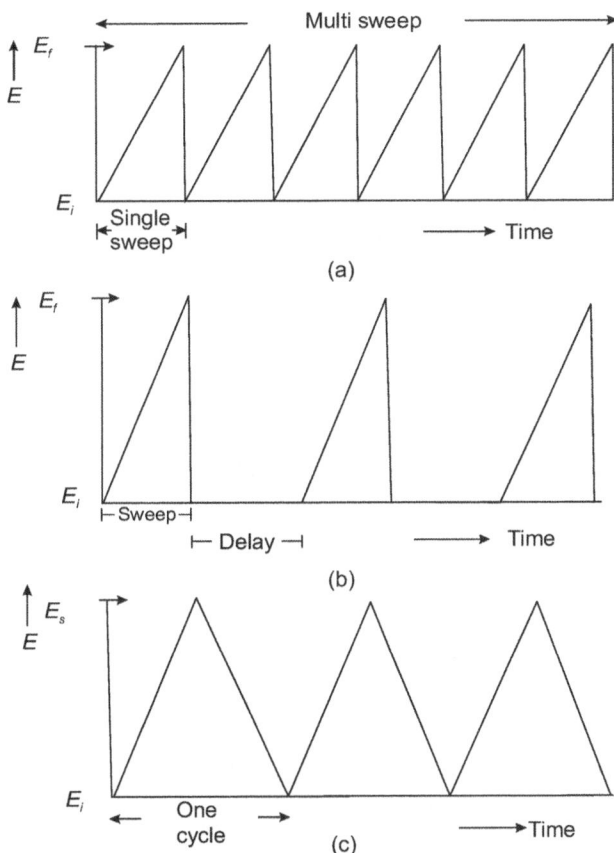

Figure 3.14   Applied waveform used in (a) Linear sweep
voltammetry (b) Linear sweep votammetry at
DME with delay period (c) Cyclic voltammetry

The potential scan direction is reversed and the potential
returned to its initial value. The material reduced in the forward
sweep, if stable, may be oxidized back to starting material on
the return sweep. Using a triangular voltage, the potential may
be continuously switched between the initial potential E_i, and
the switching potential E_s, to give the electrochemical technique
of cyclic voltammetry. Figure 3.14 shows diagrammatically the

potential as a function of time in linear sweep and in cyclic voltammetry.

THEORY FOR FARADAIC PROCESSES

Figure 3.15 shows an LSV voltammogram recorded for an unstirred solution of electroactive species at capillary electrode (DME). The peak, rather than sigmoidal shape of the LSV, is explained by depletion terms occurring in the mass transfer. The peak height is a function of concentration.

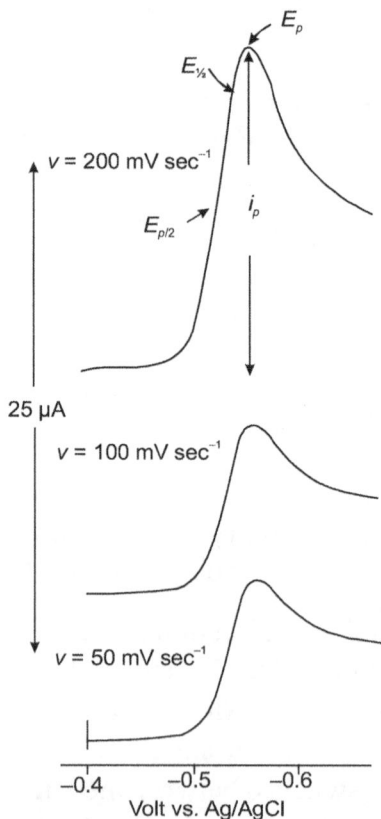

Figure 3.15   Linear sweep voltammogram

When potential is applied, current starts to flow as the reduction potential of the electroactive species is reached. In polarography, the initial conditions are regenerated by the drop growth–drop fall sequence and electrolysis which commences afresh at each potential to establish the concentration gradient. In this technique, the electrode surface is not renewed and events occurring at preceding potentials must be accounted for. Thus, depletion of the electroactive species in the vicinity of the electrode surface soon occurs as the duration of electrolysis increases, and the concentration gradient and current become smaller. At potentials corresponding to the limiting current region in DC polarography, the LSV experiments can be equivalent to a normal electrolysis in which the current would decay to zero if all the electroactive materials were to be reduced.

It is therefore easy to account for the peak shape and to recognize that the peak current i_p must depend not only on the concentration but also on n, the number of electrons transferred to the surface area A of the electrode, the diffusion coefficient D and the scan rate (v). The peak current is greater than the diffusion current in DC. The faster the scan rate, the larger the current because the diffusion layer has less time to increase in thickness and the dependence of i_p on v is explained.

If the product of the electrode process is stable, then on reversing the scan direction this material can be oxidized back to the starting material. Figure 3.16 shows a cyclic voltammogram for a chemically reversible system. A chemically reversible system is characterized by equal peak height for the reduction and oxidation processes. If the product is unstable and reacts before the reverse scan takes place, then no wave will be seen on the reverse scan.

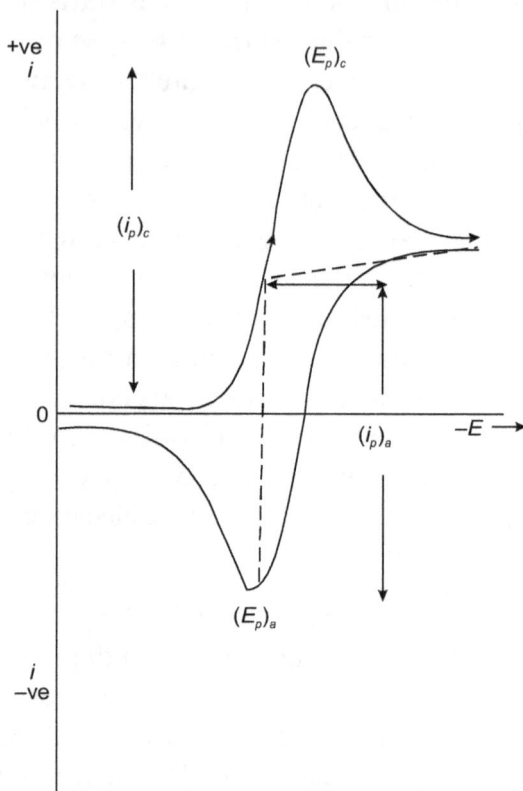

Figure 3.16   Cyclic voltammogram for a reversible process

Reversible charge transfer   For a reversible process,

$$A + ne^- \rightleftharpoons B$$

taking place at a planar electrode where A and B are soluble, the current corresponding to peak i_p is given by Randles–Sevick equation.

$$i_p = 2.69 \times 10^5 \, n^{3/2} A D^{1/2} v^{1/2} C_0 z$$

where,

n is the number of electrons transferred

A is the area of the electrode

D is the diffusion coefficient

$i_p/2$ is the scan rate in mV/s

C_0 is the concentration of electroactive species.

The peak obtained is somewhat broad, so the peak potential may be difficult to determine. It is sometimes more convenient to report the potential at $1/2\,i_p$, called the half-peak potential, $E_{1/2}$ which is

$$E_{p/2} = E_{1/2} + \frac{28.0}{n} \text{ mV at } 25°C$$

or

$$E_p - E_{p/2} = \frac{0.057}{n}$$

$E_{p/2}$ precedes $E_{1/2}$ by $\dfrac{28.0}{n}$ mV.

In cyclic voltammetry, the peak separation between cathodic (E_p,c) and anodic scan (E_p,a) will be $(E_p,c)-(E_p,a) = \dfrac{57}{n}$ mV at 25°C. The characteristics for a reversible process are

i. E_p is independent of scan rate

ii. i_p vs $v^{1/2}$ is linear passing through origin

iii. i_p is dependent on concentration

iv. $\dfrac{i_{p,a}}{i_{p,c}} = 1$ for all scan rates.

v. $E_{p,c} - E_{p,a} = \dfrac{57}{n}$ mV

All equations and discussions on cyclic voltammetry assume that the switching potential is substantially more negative than $E_{p,c}$ for reduction or substantially more positive than $E_{p,a}$ for oxidation.

Non-reversible charge transfer For a totally irreversible reduction taking place at a plane electrode,

$$O + ne^- \xrightarrow{\;k_s\;} R$$

The current is given by

$$i_{irr} = n(n\alpha)^{1/2} A D^{1/2} v^{1/2} C_0$$

There is no peak obtained during reverse scan

$$E_p - E_{p/2} = \dfrac{0.048}{\alpha n}$$

The diagnostic criteria for an irreversible process are the following.

i. i_p is a function of $v^{1/2}$

ii. E_p is dependent on concentration

iii. E_p is dependent on α, k_s and v

iv. There is a cathodic shift in the peak potential or half-peak potential of about $\dfrac{30}{n\alpha}$ mV for each tenfold increase in v

v. $\dfrac{i_{p,a}}{i_{p,c}} = 0$ for all v

For the quasi-reversible process, the separation of cathodic and anodic peak potentials $(E_{p,c} - E_{p,a})$ is a function of k_s and scan rate. For a given scan rate, smaller the k_s, the larger the separation in the peak potentials and vice versa.

COUPLED CHEMICAL REACTIONS

If a homogeneous chemical reaction is coupled to the charge transfer reaction, stationary electrode polarography provides an extremely powerful method of investigating the kinetic parameters. Several of the important kinetic parameters are discussed, including those which involve first-order reactions preceding, following catalytic chemical reactions.

Chemical reaction preceding a reversible charge transfer A large number of coupled chemical reactions involves cases in which the electroactive species is produced by a homogeneous first-order chemical reaction preceding a reversible charge transfer

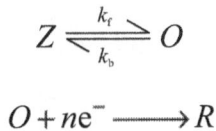

$$ Z \underset{k_b}{\overset{k_f}{\rightleftharpoons}} O $$

$$ O + ne^- \longrightarrow R $$

only the case in which the chemical reaction is reversible is relevant, however, there are no restrictions on possible values of the equilibrium constant. The diagnostic criteria for a $C_r E_r$ mechanism are

    i. An anodic shift in potential with an increasing scan rate.

    ii. The anodic portion on the reverse scan is not affected as much as the forward scan.

iii. An anodic shift in the peak potential by $\dfrac{29}{n}$ mV for every tenfold increase in v.

iv. Plot of $\dfrac{i_p}{v^{1/2}}$ vs v decreases with increasing v.

v. i_p is nonlinear with concentration. $\dfrac{i_p}{c}$ is less than reversible case.

vi. $\dfrac{i_{p,a}}{i_{p,c}}$ increases with increasing v.

Chemical reaction preceding an irreversible charge transfer   For a first-order chemical reaction preceding an irreversible charge transfer,

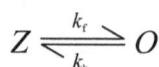

$$Z \underset{k_b}{\overset{k_f}{\rightleftharpoons}} O$$

$$O + ne^- \rightarrow R$$

The stationary electrode polarograms are qualitatively similar to chemical reaction preceding a reversible charge transfer, except that no anodic current is observed in cyclic voltammetry and further, the curves are even more drawn out because of the effect of the electron transfer coefficients.

Charge transfer followed by a reversible chemical reaction   For a reversible chemical reaction following a reversible charge transfer

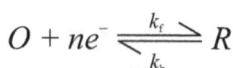

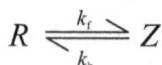

$$O + ne^- \underset{k_b}{\overset{k_f}{\rightleftharpoons}} R$$

$$R \underset{k_b}{\overset{k_f}{\rightleftharpoons}} Z$$

If the rate of the chemical reaction is very fast, the system will be in equilibrium at all times, and the only effect will be an anodic displacement of the wave along the potential axis. If the chemical reaction is very slow, so that essentially no chemical reaction takes place during the experiment, the curve should again be the normal reversible shape, but should appear at its normal potential.

If the charge transfer is irreversible, a succeeding chemical reaction will have an effect on the stationary electrode polarogram. The chemical reaction can still be studied if either substance R or Z is electroactive at some other potential but such cases will be easily handled by step functional controlled potential electrolysis.

Charge transfer followed by an irreversible chemical reaction   Consider an irreversible chemical reaction following a reversible charge transfer reaction

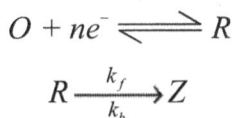

$$O + ne^- \rightleftharpoons R$$

$$R \xrightarrow[k_b]{k_f} Z$$

For small values of kinetic parameter, the chemical reaction has a little effect, and a reversible stationary electrode polarogram is obtained at its normal potential. For large values of kinetic parameter, no current is observed on scan reversal and the shape of the curve is similar to that of totally irreversible charge transfer. The diagnostic criteria for the mechanism is as follows.

i. E_p is more positive than reversible case.

ii. There is a cathodic shift in the potential by about $\dfrac{30}{n}$ mV for every tenfold increase in v.

iii. Dependence of I_p on v approaches reversible case with increasing v.

iv. i_p depends linearly on C. $\dfrac{i_p}{c}$ is slightly greater than reversible case.

v. $i_{p,c}$ is greater than $i_{p,a}$ at low scan rates.

vi. $i_{p,a}/i_{p,c}$ is unity with increasing v.

Catalytic reaction with reversible charge transfer

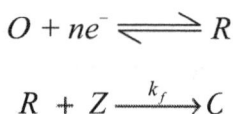

$$O + ne^- \rightleftharpoons R$$

$$R + Z \xrightarrow{k_f} C$$

The catalytic reaction scheme almost always involves a non-electroactive species (Z) in the following chemical reaction which regenerates starting material. The effect of the chemical reaction on the cathodic portion of the cyclic wave would be an increase in the maximum current. The characteristics of a catalytic wave are as follows:

i. There is an anodic shift in peak potential with increasing v.

ii. i_p dependent on v approaches reversible case with increasing v.

iii. i_p vs C is nonlinear. $\dfrac{i_p}{c}$ is less than the reversible case.

iv. $\dfrac{i_{p,a}}{i_{p,c}} \sim 1$ for all v.

v. Half-peak potential, $E_{p/2}$ is independent of v at both low and high values of v.

vi. $\dfrac{\Delta E_{p/2}}{\Delta \log v}$ is maximum in the kinetic region.

Adsorption reaction If the product or reactant is strongly adsorbed, a separate adsorption peak may occur prior to or after the normal peak. The following criteria can be applied to test for adsorption.

i. Adsorption-controlled waves are frequently symmetrical about i_p, unlike a normal wave.

ii. $\dfrac{i_p}{Cv^{1/2}}$ increases rapidly with increasing scan rate; however, $\dfrac{i_p}{Cv}$ may remain constant.

iii. $\dfrac{i_p}{C}$ increases with decreasing concentration.

Diagnostic criteria for adsorption

Reactant adsorbed

i. E_p shifts cathodically with increasing v.

ii. i_p vs C is non-linear.

iii. i_p increases with increasing v.

iv. $\dfrac{i_{p,a}}{i_{p,c}} \leq 1$ approaches 1 at low scan rates.

Product adsorbed

i. E_p shifts anodically with increasing v.

ii. i_p vs C is non linear.

iii. i_p decreases with increasing v.

iv. $\dfrac{i_{p,a}}{i_{p,c}} \geq 1$ approaches 1 at low scan rates.

STRIPPING VOLTAMMETRY

Stripping voltammetry is an electroanalytical technique which generates its extremely favourable faradaic to charging current rates. The extremely large faradaic currents per unit concentration give rise to extremely low detection limits. The DC potential ramp in cyclic voltammetry and AC polarography is a triangular voltage, where reduction of electroactive species in solution may occur when the negative potential direction sweep. When the scan direction is reversed, oxidation of stable product generated at the electrode surface may occur, and a complete cyclic voltammogram may include cathodic and anodic current components. If a reversible process of the $A + ne \rightleftharpoons B$ type occurs, the peak heights of the reduction and oxidation processes are equal. This will occur only when B is soluble in solution. If B forms an amalgam so that the electrode process is of the kind,

$$A + ne \rightleftharpoons BCHO$$

and the reduction is carried out at a relatively small volume of hanging mercury drop electrode (HMDE), it is clear that during a reverse scan where the metal is stripping from the amalgam and oxidized back into the solution, the peak height is larger than that of all reduction processes. The reverse sweep of a cyclic voltammogram for systems involving amalgam formation at an HMDE, therefore shows the analytically desired enchancement of the faradaic current and constitutes a special form of anodic stripping voltammetry.

ANODIC STRIPPING VOLTAMMETRY

Anodic stripping voltammetry has two steps in common. First, the analyte species in the sample solution is concentrated on to or into a working electrode. It is this preconcentration step that results in the exceptional sensitivity that can be achieved. During the second step, the preconcentrated analyte is measured or stripped from the electrode by the application of a potential scan. Any number of potential wave forms can be used for the stripping step (i.e., differential pulse, square wave, linear sweep or staircase). The most common are differential pulse and square wave due to the discrimination against the charging current. However, square wave has the added advantages of faster scan rates and increased sensitivity relative to differential pulse.

Consider a conventional DC polarographic system set-up with an oxygen-free solution of supporting electrolyte containing one or more tons of reducible ions at a Hg electrode (HMDE). The potentiostat of the polarograph is set to a fixed value which is 0.2–0.4 V more negative than the highest reduction potential encountered among the reducible ions; then electrolysis will occur, deposition of metals will take place on the HMDE cathode and usually amalgam formation will take place. The rate of amalgam formation will be governed by the magnitude of the current flowing, by the concentration of the reducible ions and by the rate at which the ions travel to the electrode; the latter can be controlled by stirring the solution. Given sufficient time, the whole of the reducible ion content of the solution may be transferred to the Hg cathode. This is called concentration step, where the metals become concentrated into the relatively small volume of the Hg drop.

The electrolysis current not stopped, the stirrer is switched off, and the cell is allowed to stand for about 30 seconds to allow the solution to become quiescent. Now the voltage sweep is reversed. The gradual increase in positive potential is applied to HMDE, which is now the anode of the cell. The current is measured and plotted against the voltage. The initial gradual increase in current corresponding to the residual current is observed. As the potential approaches the oxidation potential of one of the metals dissolved in mercury, ions of that metal oxidize or pass into the solution, and the current increases rapidly and attains a maximum value when the potential approaches oxidation potential. The metal is said to be "stripped" from the amalgam. Then the current declines from its maximum value and settles down to a new steady value. The curve shows a peak. A typical stripping voltammogram is shown in Figure 3.17.

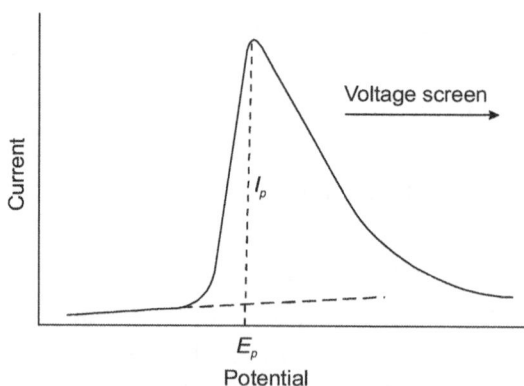

Figure 3.17   Stripping voltammogram

The peaks are characterized by the peak potential, E_p, by the peak current, I_p, and by the breadth b. These parameters are dependent upon characteristics of the electrode and upon the rate of the voltage sweep during the stripping process.

The magnitude of the peak current is proportional to the concentration in the amalgam of the metal being stripped and is therefore proportional to its concentration in the original solution.

The concentration step is carried out for a definite time under reproducible conditions (the solution may be stirred or the electrode rotated at constant speed, e.g. to improve the efficiency of electrolysis).

ELECTRODES

The electrochemical cell and instrumentation used in stripping voltammetry is usually similar to that for polarography. Thus, a three-electrode potentiostat system is employed to minimize the *iR* drop effects with suitable working, auxiliary and reference electrodes.

The HMDE is one of the most popular working electrodes for stripping voltammetry. The commercially available HMDE consists of a microsyringe with a micrometer to control the size of the drop. HMDE offers simplicity, economy and reproducibility. Alternatively mercury drops may be suspended from the tip of a platinum wire imbedded in a glass rod. Two basic disadvantages are noticed with HMDE. First it has a low surface area-to-volume ratio. The relatively small surface area reduces the efficiency of the pre-electrolysis step. Second, only low stirring rates of the solution may be employed with HMDE to avoid dislodging of the drop.

A number of different types of carbon electrodes (solid electrodes) have been reported. But poor results are obtained by depositing samples directly on to a solid metal or carbon electrodes. Problems associated with surface contamination and reproducibility are often encountered.

Theory and Techniques

A rigorous theory for the controlled-potential electrolysis step is difficult to implement because the mass transport processes are complicated and not always extremely well known or controllable in actual practice. Since the controlled-potential electrolysis step in anodic stripping voltammetry is performed at a potential of 300 to 400 mV more negative than the polarographic half-wave potential, and the solution is stirred or the electrode rotated, the current flow at time t is given by Levich equation.

$$i(t) = k_1 nFAD^{2/3} \omega^{1/3} v^{1/6} C_A$$

where,

> k_1 is a constant for the electrode
>
> ω is the rate of electrode rotation or solution stirring
>
> v is the kinematic viscosity of the solution, and
>
> C_A is the concentration of the metal ion in solution at deposition time.

The viscosity of aqueous solutions does not vary greatly. However, the stirring rate of the solution and the rotation rate of the electrodes are the important variables to be considered. The rate of rotation can be increased to a point where the ability to maintain a drop at a hanging mercury drop electrode becomes critical.

The peak potential E_p of the stripping curve at HMDE is given by the usual expression for oxidation in linear sweep voltammetry

$$E_p = E_{1/2} + \frac{1.1 RT}{nF}$$

where $E_{1/2}$ is the reversible half-wave potential.

Table 3.1   Relative sensitivity of some electrochemical techniques

Techniques	Limits of detection for Pb(II)
Ion-selective electrode	10^{-5} M
DC polarography at DME	10^{-6} M
Differential pulse polarography at SMDE	10^{-7} M
Differential pulse ASV at HMDE	10^{-10} M
DC ASV at mercury film	10^{-11} M
Square-wave ASV at mercury film	10^{-12} M

CATHODIC STRIPPING VOLTAMMETRY

CSV can be used to determine substances that form insoluble salts with the mercurous ion. Application of relatively positive potential to the mercury electrode in a solution containing such substances results in the formation of an insoluble film on the surface of the mercury electrode. A potential scan in the negative direction will then reduce (strip) the deposited film into solution. This method has been used to determine inorganic anions such as halides, selenide and sulphide and oxyanions such as MoO_4^{2-} and VO_3^{5-}. In addition, many organic compounds such as nucleic acid bases also form insoluble mercury salts and may be determined by CSV.

ADSORPTIVE STRIPPING VOLTAMMETRY

AdSV is quite similar to anodic and cathodic stripping methods. The primary difference is that the preconcentration step of the analyte is accomplished by adsorption on the electrode surface or by specific reactions at chemically modified electrodes rather than accumulation by electrolysis. Many organic species (such as haeme, chlorpromazine, codeine, and cocaine) have been determined at micromolar and nanomolar concentration levels using AdSV; inorganic species have also been determined. The adsorbed species is quantified by using a voltammetric technique such as DPV or SWV in either direction (negative or positive) to give a peak-shaped voltammetric response with amplitude proportional to concentration.

Problems

1. Calculate the current for diffusion-controlled electrolysis at a spherical electrode under the conditions; $n = 1$, $C^* = 1.00$ mM, $D = 10^{-5}$ cm^2/sec at $t = 0.1, 0.5, 1, 2, 3, 5, 10$ sec and as $t \rightarrow a$. Plot both i vs t curves on the same graph.

2. The following measurements were made on a reversible polarographic wave at 25°C. The process could be written as $O + ne^- \rightarrow R$

E (V vs SCE)	i (μA)
−0.395	0.48
−0.406	0.97
−0.415	1.46
−0.422	1.94
−0.431	2.43
−0.445	2.92

Calculate:

a. The number of electrons involved in the electrode reaction.

b. The formal potential (vs NHE) of the couple involved in the electrode reaction, assuming $D_0 = D_R$.

3. Derive the polarographic current-potential for the reduction of a simple metal ion to a metal that is insoluble in mercury. Assume that the electrode reaction is

$$M^{n+} + ne^- \rightarrow M$$

Also assume that the electrode reaction is reversible, and that the activity of solid M is constant and equal to 1. How does $E_{1/2}$ vary with I_d and with the concentration of M^{n+} ?

4. The following measurements were made at 25°C on the reversible wave for the reduction of a metallic complex ion to metal amalgam:

[Ligand], M	$E_{1/2}$ (V vs SCE)
0.10	−0.448
0.50	−0.531
1.0	−0.566

a. Calculate the number of ligands associated with the metal M^{n+} in the complex.

b. Calculate the instability constant of the complex, if $E_{1/2}$ for the reversible reduction of the simple metal ion is +0.081 V vs. SCE. Assume that the D values for the complex ion and the metal ion are equal, and that all activity coefficients are unity.

5. a. Reduction of many organic substances involve the hydrogen ion. Derive the polarographic equation for the reversible reaction

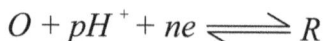

$$O + pH^+ + ne \rightleftharpoons R$$

where both O and R are soluble substances, and only O is initially present in solution at a concentration C_o^*.

b. What experimental procedure would be useful for determining R?

6. A polarogram of molecular oxygen in air-saturated 0.1 M KNO_3 showed two peaks at −0.4 V and −1.7 V. The concentration of O_2 is about 0.25 mM. At E = −0.4 V vs. SCE, $(i_d)_{max} = 3.9\ \mu A$, $t_{max} = 3.8$ sec, and $m = 1.85$ mg/sec. At $E = -1.7$ V vs. SCE, $(i_d)_{max} = 6.5\ \mu A$, $t_{max} = 3.0$ sec, and $m = 1.85$ mg/sec. Calculate $(I)_{max}$ at each potential. Is the ratio of the two values what you expect? Explain any discrepancy in chemical terms. Calculate the diffusion coefficient for O_2 using the more appropriate constant.

7. The oxidation of o-dianisidine (o-DIA) occurs in a Nernstian 2e reaction. For a 2.27 mM solution of (o-DIA) in 2 M H_2SO_4 at a carbon paste electrode of area 2.73 mm² with a scan rate of 0.500 V/min, $I_p = 8.19$ mA. Calculate the D value for o-DIA. What I_p is expected for $v = 100$ mV/sec? What I_p will be obtained for $v = 50$ mV/sec and 8.2 mM o-DIA?

8. The following figure shows a cyclic voltammogram taken for a solution containing benzophenone (BP) and tri-*p*-tolyamine (TPTA), both at 1 mM in acetonitrile. Benzophenone can be reduced inside the working range if acetonitrile and TPTA can be oxidized. However, benzophenone cannot be oxidized, and TPTA cannot be reduced. The scan shown here begins at 0.0 V vs QRE and first moves toward positive potentials. Account for the shape of the voltammogram.

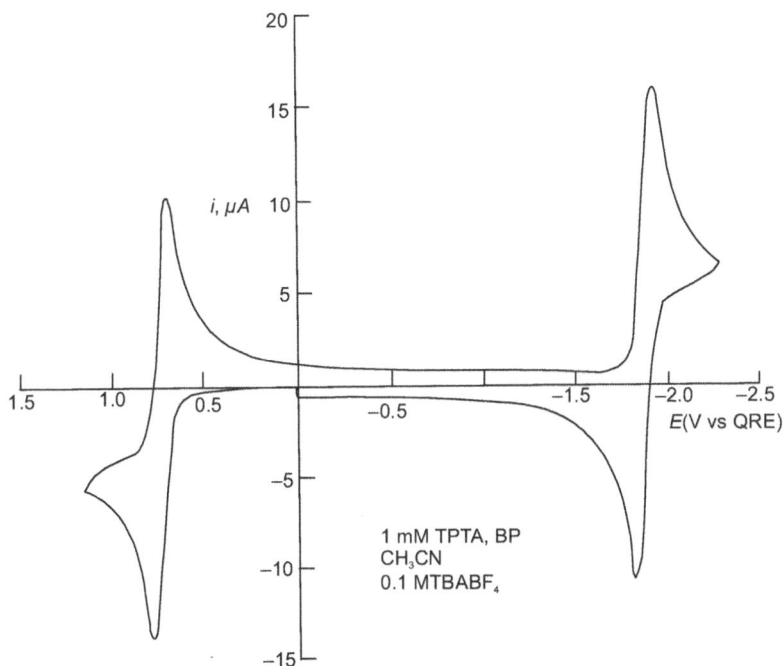

1 mM TPTA, BP
CH$_3$CN
0.1 MTBABF$_4$

9. Consider the electrochemical reduction of molecular oxygen in an aprotic solvent such as a pyridine or acetonitrile. In general, a cyclic voltammogram like that in the following figure is obtained. The polarogram (at a DME) gives a linear plot of E vs log $[(i_d - i)/i]$ with a slope of 63 mV. The reduction product at -1.0 V vs SCE gives an ESR signal. If methanol is added in small quantities, the cyclic voltammogram shifts toward positive potentials, the forward peak rises in magnitude, and the reverse peak disappears. These trends continue with increasing methanol concentration until a limit is reached with reduction near -0.4 V vs SCE. The polarogram under these limiting conditions is approximately twice as high as it was in methanol-free solution and the wave slope is 78 mV.

    (a) Identify the reduction product in methanol-free solution.

(b) Identify the reduction product under limiting conditions in methanol-containing solution.

(c) Comment on the charge transfer kinetics in methanol-free solution.

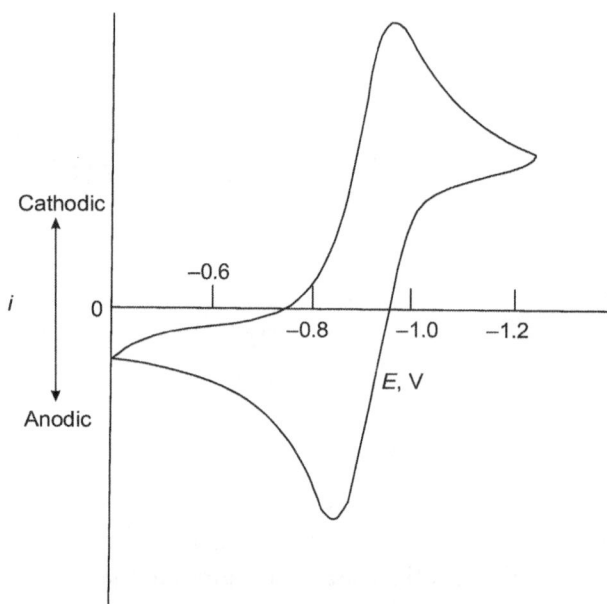

10. The cyclic voltammetry of azotoluene was studied under the following conditions.

Solutions

N,N^1-dimethylformamide containing 0.10 M tetra-*n*-butylammonium perchlorate as supporting electrolyte and 0.68 mM in azotoluene; working electrode: planar platinum disc, 1.54 mm²; reference electrode: SCE; temperature: 25°C. A typical cyclic voltammogram is shown in the following figure and the data obtained are as follows:

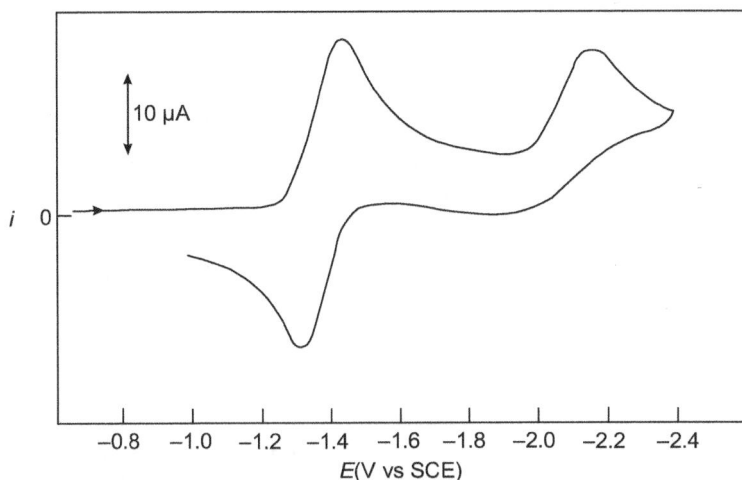

Scan rate, (mV/sec)	First wave				Second wave		
	$i_{p,c}$ (μA)	$i_{p,a}$ (μA)	$-E_{p,c}$ (V vs SCE)	$-E_{p,a}$ (V vs SCE)	$i_{p,c}$ (μA)	$-E_{p,c}$ (V vs SCE)	$-E_{p,a}$ (V vs SCE)
430	8.0	8.0	1.42	1.36	7.0	2.10	2.00
298	6.7	6.7	1.42	1.36	6.5	2.09	2.00
203	5.2	5.2	1.42	1.36	4.7	2.08	2.00
91	3.4	8.4	1.42	1.36	3.0	2.07	1.99
73	3.0	2.9	1.42	1.36	2.8	2.06	1.98

Coulometry shows that the first reduction step involves one electron. Work out this set of data and discuss what information is obtained about the reversibility of the reaction, stability of products, diffusion coefficients, etc.

11. Suppose bromide ion is to be determined at very low concentrations. This is done by depositing bromide on a silver electrode, which is held at a potential where the following reaction occurs:

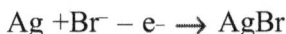

$$Ag + Br^- - e- \longrightarrow AgBr$$

(A typical deposition potential is +0.2 V vs SCE). Stripping is carried out by scanning in a negative direction to reverse the deposition. In general, it is observed that the response during stripping shows a complex dependence on deposition time. Explain this effect. What problems would be present in quantitative analysis?

12. a. A study of seawater by stripping analysis reveals an anodic copper peak having a height of 0.13 mA when deposition is carried out at –0.5 V.However, deposition at –1.0 V yields a larger peak of 0.31 mA. Account for these results.

   b. Standard addition of 10^{-7} M Cu^{2+} elevates the peaks in both cases by 0.24 mA. Comment on the feasibility of obtaining polarograms of any types on this solution. What responses would you get for DC, normal pulse, and differential pulse experiments? Would any of these be useful analytical information?

13. An analysis for lead at the HMDE gives rise to a peak current of 1 mA under condition in which the deposition time is held constant at 5 min and the sweep rate is 50 mV/sec.What currents would be observed for sweep rates of 25 and 100 mV/sec?

14. Consider a chronopotentiometric experiment dealing with two components that are reversibly reduced in waves separated by 500 mV. Derive an expression for the second transition time in an experiment carried out in a thin-layer cell. Compare and contrast the properties of multicomponent systems in thin-layer chronopotentiometry with those of the semi-infinite method.

References

Bard, A.J. and Faulkner, L.R. (1980). *Electrochemical Methods*. Wiley, New York.

Gosser, D.K. (1993). *Cyclic Voltammetry: Simulation and Analysis of Reaction Mechanism*. VCH Publishers, New York.

Kissinger, P.T. and Heineman, W.R. (1984).*Laboratory Techniques in Electroanalytical Chemistry*. Marcel Dekker, New York.

Kissinger, P.T. and Heineman,W.R., (1983). "Cyclic voltammetry,"*J. Chem. Ed.* 60, 702.

Nicholson, R.S. and Shain, J. (1964). *Anal. Chem.* 36, 706.

Osteryoung, J. and Osteryoung, R.A. (1985). "Square wave voltammetry." *Analytical Chemistry*. 57, 101A.

Wang, J. (1985). *Stripping Analysis*. VCH Publishers, Deerfield Beach, FL.

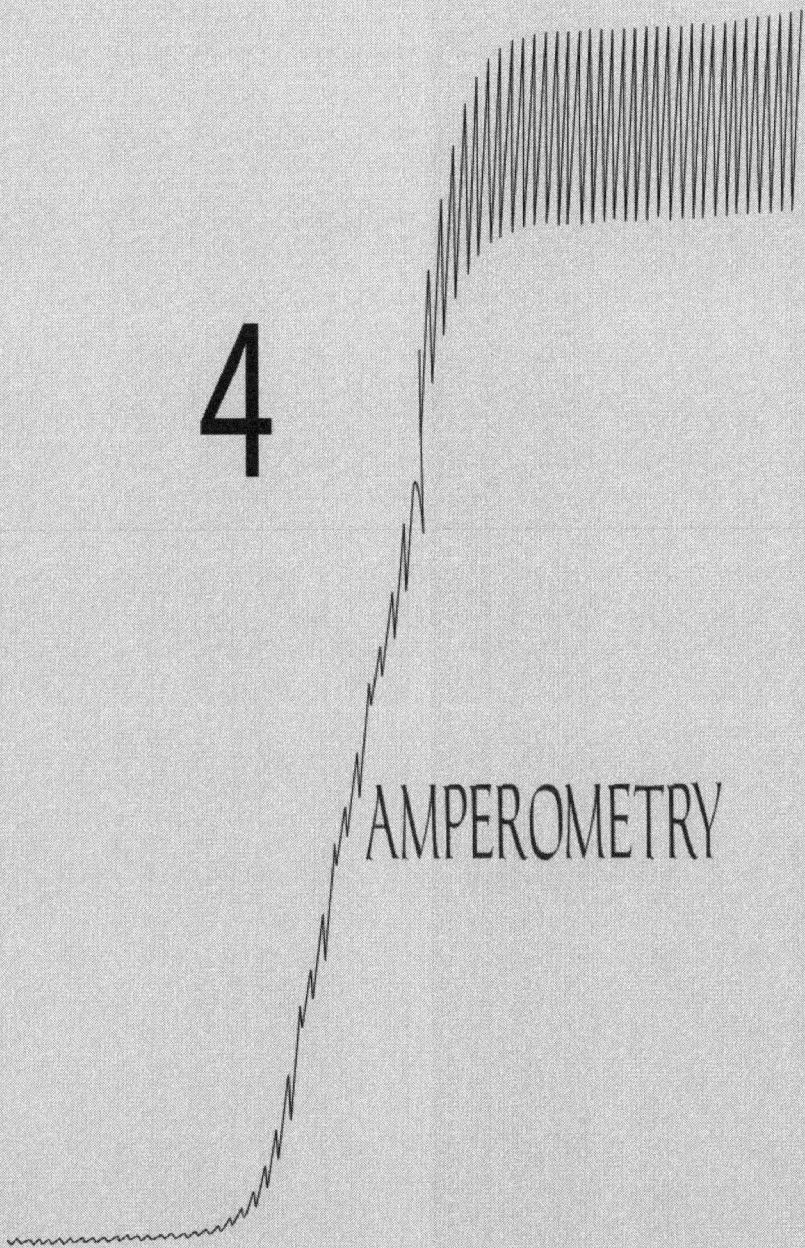

4

AMPEROMETRY

Amperometry is a technique where current is measured when constant potential is applied as a function of time or volume. Oxidation or reduction reaction is forced to occur on the surface of electrode, by selecting an applied potential. The resultant current is measured as a function of time or volume of titrant. This simplest technique finds application in many fields, namely development of biosensors, and in environmental analysis.

PRINCIPLE

The application of constant potential on the electrode surface results in oxidation or reduction reaction. The current is measured as a function of time. It is important to recognize that electrochemical detection is a surface technique, which means molecules not adjacent to the electrode must be moved to the surface to react. Most amperometric techniques are based on thin layer hydrodynamic chronoamperometry, which is the measurement of current at controlled potential as a function of time in a stirred solution. The electrode is placed in a flowing stream (hydrodynamic) configured as a thin film. The thickness of the film is varied from 15 to 125 mm.

In a solution having sufficient electrolyte concentration, the potential is applied across a very thin interfacial region between the working electrode surface and the bulk solution. The electric field in this zone is therefore very large of the order of 10^5 to 10^6 V/cm.

All amperometric determinations depend on Faraday's law. The Faraday's law is given by

$$Q = nFN$$

where,

> Q is the number of coulombs,
>
> N is the mole of electroactive material,
>
> n is the number of electron equivalents that are lost or gained in the transfer process per mole of material, and
>
> F is the Faraday's constant.

Differentiation of the above equation with respect to time (t) gives current (I) which is the measure of the rate at which the material is oxidized or reduced.

$$\frac{dQ}{dt} = I = nFA\frac{dN}{dt}$$

When sufficient potential is applied, the reduction or oxidation is highly favoured, which results in all electrochemically active materials that come into contact with the electrode being converted to product. Under these conditions, the current depends on mass transport. The rate of mass transport is determined by diffusion coefficient, D, and concentration gradient, $\frac{dc}{dx}$ (the change in concentration of electroactive species with respect to distance). The rate of mass transport is given by

$$\frac{dN}{dt} = -D\left(\frac{dc}{dx}\right)_{x=0}$$

or in terms of the current response.

$$I = -nFAD\left(\frac{dc}{dx}\right)_{x=0}$$

Under constant flow or controlled hydrodynamic conditions, the concentration gradient is constant because the diffusion layer, ∂, is nonvarying:

$$I = nFA \frac{DC_0^*}{\partial}$$

where C_0^* is the unperturbed concentration of the reactant. The above equation relates the current and the concentration of reactant passing through an electrochemical transducer cell.

This technique has an advantage over most analytical techniques, which involves a direct conversion of chemical information to an electrical signal without the use of optical or magnetic carriers. If a reduction takes place, electrons flow from the electrode to the molecule by a heterogeneous transfer; and in oxidation, transfer of electrons is in the opposite direction. Under steady-state conditions, the current measured is contributed from these sources (i) the background electrolyte, (ii) the electrode material and (iii) the analyte. The medium and the electrodes are chosen so that contributions of the first two sources are as small as possible and the small residual current that emerges from these sources is electronically removed before quantitation of the analyte.

In this experiment, initially the energy applied to the cell at low positive or negative potential is insufficient to cause any reaction to occur. As the applied potential increases in the positive or negative direction, the energy requirement for the reaction is partially met, and a faradaic current increases. If the potential is further increased, the faradaic current increases until a potential is reached. This current vs potential plot is called hydrodynamic voltammogram (HDV).

Electroactive functional groups have characteristic redox potential. Hence, the measurement of peak potential may be used as an approximate indication of the voltage required from an amperometric detector. The preferred method of determining the peak potential is HDV.

INSTRUMENTATION

The basic components of the amperometric set-up are given in Figure 4.1.

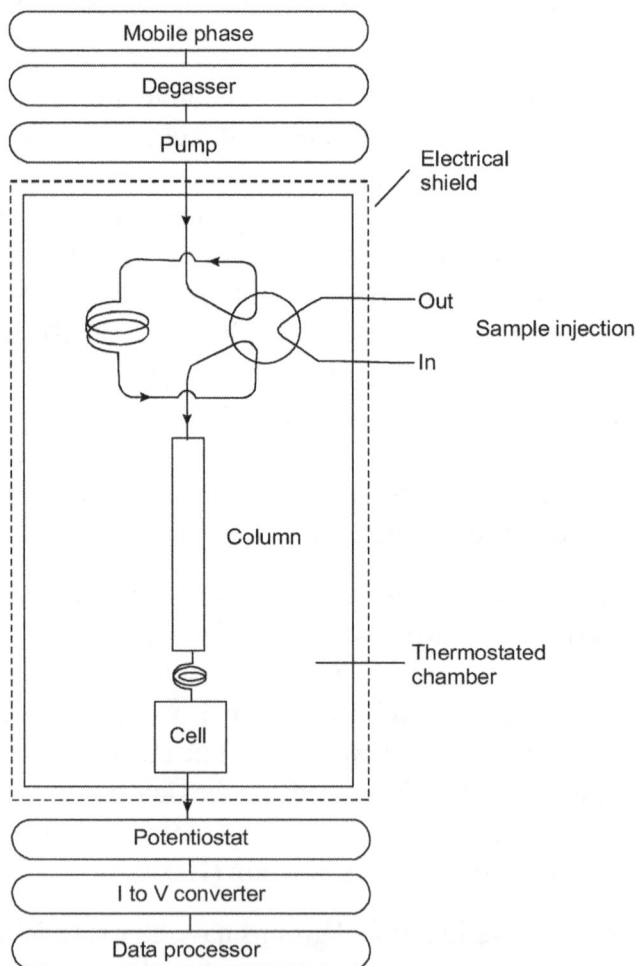

Figure 4.1   Basic components of amperometric set-up

Pump It provides a constant flow of mobile phase, the electrolyte, to an injection valve where the sample is introduced. The flow should be as pulseless as possible to minimize baseline noise. Amperometric detectors respond to pressure pulsation. Dual-piston pumps with a pulse dampener are most commonly used. Because amperometry involves a surface reaction between the electrode and the mobile phase and analyte, it is not surprising that the composition of the mobile phase plays an essential role in a successful experiment. It requires a low electrochemical activity and an electrolyte of 0.01 to 0.1 M ionic strength to minimize ionic resistance.

Degasser Gases dissolved in mobile phase can contribute to baseline noise and poor performance. It is desirable that the pump receives mobile phase with gases below saturation, minimizing the chance of air bubbles formed during the pump refill stroke or in the detector cell.

Injector valve Sample injection may be manual or automated but should be of short duration. Because the detector responds to pressure changes, it is affected by the movement of the injector valve from the load to inject position. In addition, there is a small pressure drop when the volume within an injection loop is compressed as it enters the pressurized mobile phase flow stream.

Analytical column The injected sample may pass through a tube directly to the amperometric cell or into a column for separation of individual components. The former situation is called flow injection analysis (FIA) and produces a single signal representing all electroactive compounds in the sample. Generally FIA is useful only if a single electroactive analyte is being determined. If the sample contains many electroactive components, an appropriate chromatographic separation is needed. The type of analytical column used depends on the application but most often consists of reversed phase materials (3 to 10 mm particles) packed into stainless steel tubing, ranging in size from 1 mm to 5 mm and ranging in length from few mm to hundreds of mm.

Transducer Cell

The thin-layer channel　It is defined by a gasket held between a stainless steel block and a polymeric block (Figure 4.2). The stainless steel block is the auxiliary electrode and provides a compartment for the reference electrode. The polymeric block contains the working electrodes. This design and, in fact, nearly all cell designs incorporate these electrodes: working, auxiliary and reference electrodes. The potential selected is applied between the reference and working electrodes while the current is passed between the auxiliary and working electrodes. Detection occurs at the working electrode in the thin-layer region. Electrodes of the same or different materials may be interchanged by simply swapping the working electrode half of the thin layer cell. Carbon paste, glassy carbon, mercury on gold, Pt and Ag have all been used. The cell volume can also be reduced to less than 300 nl by simple gasket changes.

The thin-layer design has consistently demonstrated some of the lowest detection limits reported because signal-to-noise increases as the electrode size decreases. The potential limitation is how small a well-sealed electrode can be made and how small a current can be measured without introducing substantial electronic noise. Electrodes having diameters of a few mm have proved to be reasonable compromises.

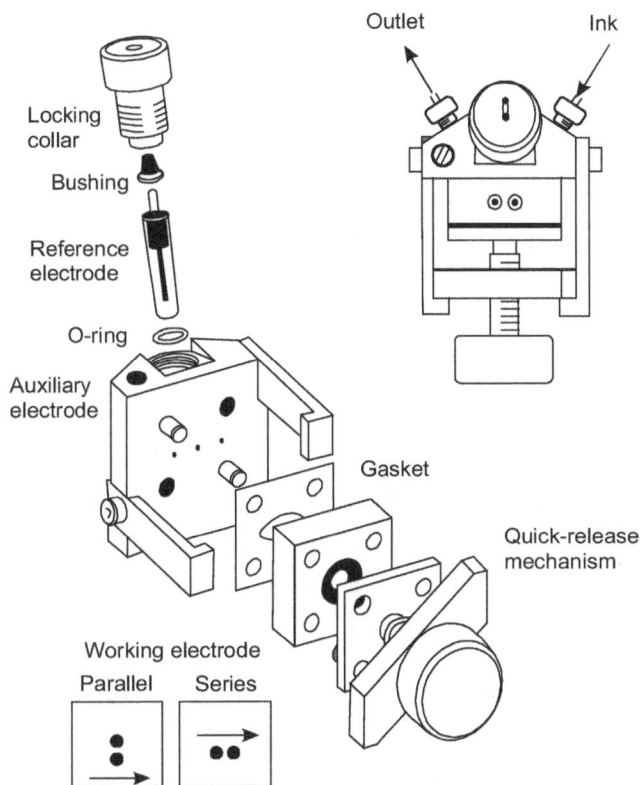

Figure 4.2   Thin layer amperometric cell

The current of an electrophoresis system gives rise to noise at the electrochemical cell and must be electrically isolated. This has been achieved through a conducting joint in the capillary tubing before the electrochemical cell. The construction of this joint as well as cell design has been varied. Extreme sensitivity has been reported, e.g. determination of the concentration of catecholamines in single nerve cell.

Data processor   The output on most amperometric detectors is an analog signal of 0 to 1.0, 0 to 0.1 or 0 to 0.01 volts.

A strip chart recorder, integrator, or computer with an A/D interface and software package can be used to collect and process the output signal.

AMPEROMETRIC DETECTION AND BIOSENSORS

The advantages of amperometric detection can be extended to the *in situ* determination of important molecules, including those of biological significance. The most widely studied application of biosensors is the determination of blood glucose.

The determination of glucose can be achieved by oxidizing glucose and monitoring it by biosensors. However, it is difficult to oxidize electrochemically because at the same potential, other blood components will also be oxidized. So it is selectively oxidized by choosing biomolecules like enzymes (e.g. glucose oxidase) or antibodies. It is very difficult to immobilize these enzymes or antibodies on the electrode as a thin film and to measure the result of interaction of glucose with the enzyme, because the reduced form of enzyme is site-specific. Hence, the cofactor related to enzyme activity (i.e., H_2O_2 produced as a result of O_2 reduction during a reaction catalysed by glucose oxidase) can be monitored and thus glucose content in blood is determined (Figure 4.3a).

Many applications using oxidase enzymes are based on this simple principle. The common problem is that H_2O_2 requires a relatively high potential at a Pt electrode, so easily oxidized materials will interfere.

Another route to link the electrode with the enzyme is via a chemical–electron shuttle, an electron transfer mediator (cofactor).

(a)

(b)

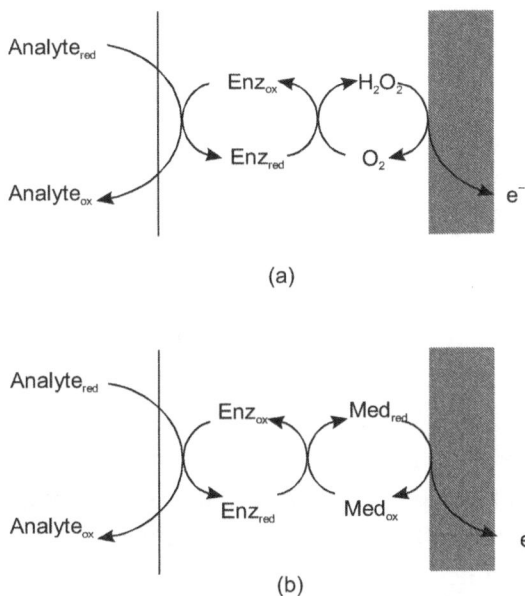

Figure 4.3 Two approaches for amperometric biosensors (a) Monitoring the hydrogen peroxide produced by an oxidase enzyme specific for the analyte of interest. (b) Monitoring a reduced enzyme cofactor (an electron transfer mediator)

The enzyme oxidizes the analyte and in turn reduces the mediator (Figure 4.3b). The reduced mediator contacts the electrode where it is oxidized and is ready to interact onceagain with the reduced enzyme. Although in principle this is very simple, in practice these devices are difficult to make and meet some commercial criteria of long-term stability, reproducibility and speed of analysis, but the benefit is sufficiently great that there is much ongoing research and development from both academic and commercial sources.

AMPEROMETRIC TITRATIONS

It is known that the limiting current is independent of the applied voltage upon a dropping mercury electrode or any other indicator microelectrodes. The factor which affects the limiting current is the rate of diffusion of electroactive material from the bulk of the solution to the electrode surface. (The migration current is suppressed by the addition of supporting electrolyte.) Hence the diffusion current is directly proportional to the concentration of the electroactive material in the solution. If some of the electroactive materials is removed by interaction with reagent, the diffusion current will decrease. This is the fundamental principle of amperometric titrations. The observed diffusion current at a suitable potential is measured as a function of volume of the titrant. The end point is the intersection of two straight lines giving the changes of current before and after the equivalence point.

If the current–voltage curves of the reagent and of the substance being titrated are not known, the polarogram must first be determined in the supporting electrolyte in which the titration is to be carried out. The potential applied at the beginning of the titration must be similar to the potential corresponding to the total diffusion current of the substance to be titrated, or of the reagent, or of both.

The amperometric titration curve and corresponding hypothetical polarograms of each individual substance are shown in Figure 4.4 where, S refers to the solute to be titrated, and R to the titrating reagent. The slight rounding off in the vicinity of the equivalence point is due to the solubility of precipitate, salt hydrolysis or dissociation of a complex. For each amperometric titration, the potential applied is chosen between the values of X and Y.

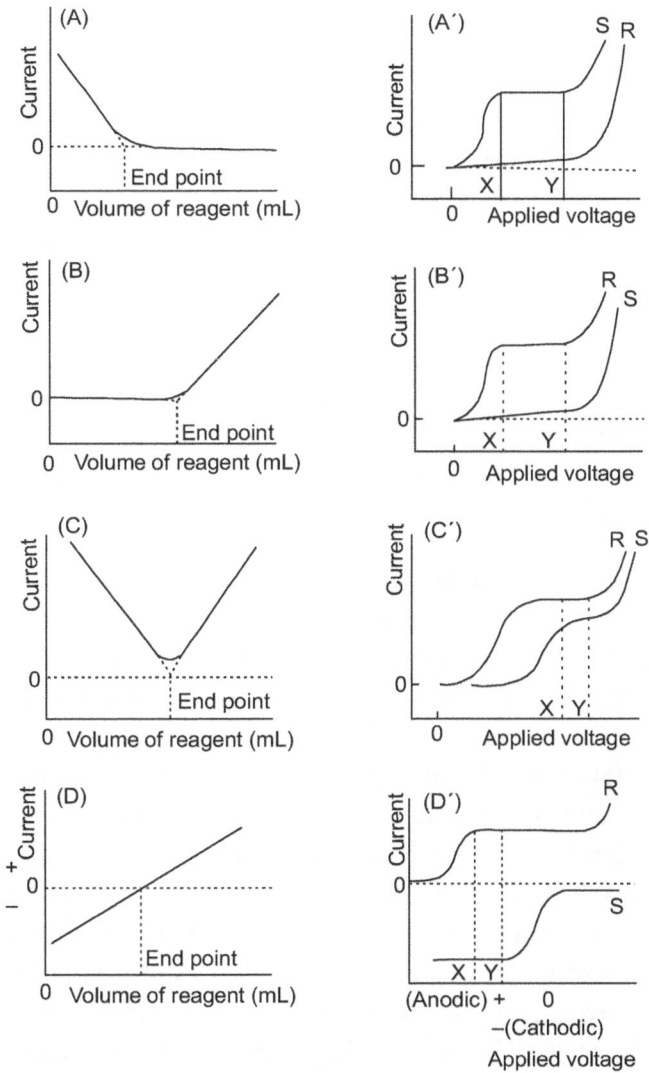

Figure 4.4   Amperometric titration curves and their hypothetical polarograms

Type A   The substance being titrated is an electroactive material and gives the polarogram, and is removed by the addition of reagent (electro-inactive material) which does not give polarogram. The substance being titrated is taken in an electrochemical cell. When potential is applied, the electroactive species gets oxidized or reduced. Hence, diffusion current increases. This current is monitored as a function of volume of electroinactive material (i.e., reagent). When a titrant is added to electroactive material in small increments, it chemically reacts with it, thereby decreasing current. The current continues to decrease till it reaches an end point. Further addition of reagent does not change the current because it is an electroinactive material. The current remains constant after end point. The intersection of two straight lines gives end point. For example, lead ions (electroactive) can be determined by the addition of oxalate or sulphate ion (inactive).

Type B   The substance being titrated gives no diffusion current whereas the reagent added to precipitate the titrant gives the polarogram. When potential is applied, there will not be any current because the substance being determined is not an electroactive material. The addition of reagent or precipitant does not change the current because it chemically reacts with the substance to be determined. When substance gets reacted completely, further addition of reagent increases the current because it is an electroactive material. The current increases as the volume of reagent increases. The intersection of two lines gives end point. For example, concentration of sulphate ions are determined by adding barium or lead ions. An electroactive precipitating reagent is added to an inactive substance.

Type C   Both the substance being titrated and the reagent give diffusion currents (electroactive). Both the substance being determined and reagent are electroactive materials.

The potential should be chosen in such a way that both the substance and reagent get reduced or oxidized. When potential is applied, there will be an appreciable current and this keeps on decreasing till the end point. After this end point, the current increases because the reagent is also an electroactive material. For example, Pb^{2+} titrated against chromate ion, Ni^{2+} against DMG (dimethyl glyoxime) and Cu^{2+} with benzoin α-oxime. The diffusion current starts decreasing before the end point and it increases after the end point since reagent is also an electroactive material. Hence a V'-shaped curve is obtained.

Type D The substance being titrated gives an anodic diffusion current and the reagent gives cathodic diffusion current. When potential is applied, there is an anodic diffusion current. The addition of reagent decreases the anodic current and reaches zero value at the end point. After the end point, the addition of reagent increases the cathodic current. Hence, the current changes from anodic to cathodic or vice versa. The end point of the titration is indicated by zero current, e.g. titration of I^- with Hg^{3+}, Cl^- with Ag^+, Ti^{3+} in an acidified tartrate medium with Fe^{3+}.

Titrations with DME

A 100-ml four-necked flat- or round-bottomed flask containing the substance to be titrated is taken. The semi-microburette, dropping electrode, a two-way gas inlet tube and agar–KCl salt bridge fitted in 4 necks. Figure 4.4 shows a titration cell. The simple electrical circuit is also shown in Figure 4.3. The voltage applied to the titration cell is supplied by two 1.5-V dry cells and is controlled by the potential divider *R*. The current flowing is read on the microammeter M.

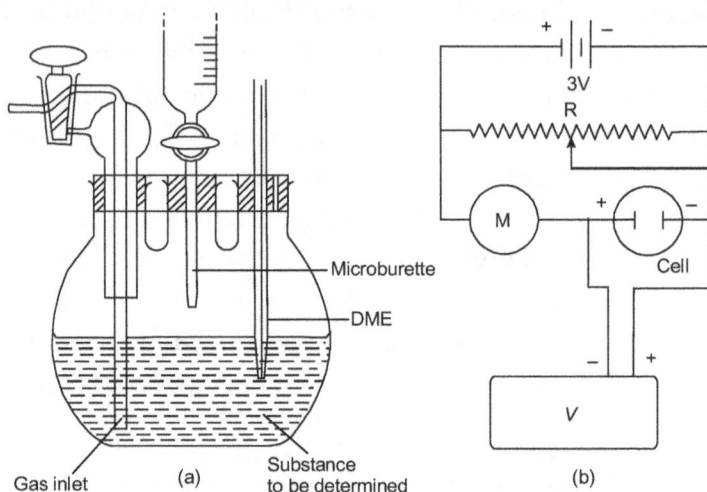

Figure 4.5   a. Amperometric titration cell b. Simple electrical circuit for amperometry

If the solute is electro-reducible, sufficient electrolyte should be added to eliminate the migration current. If the reagent is electro-reducible and the solute is not, the addition of supporting electrolyte is not required, since sufficient electrolyte is formed during the titration to eliminate the migration current beyond the end point.

A known volume of the solution under investigation is placed in the titration cell. The dissolved O_2 is removed by passing a slow stream of N_2 and the initial diffusion current is noted. A known volume of the reagent (concentration is 10 to 20 times higher than that of the solute) is added from a semi-microburette. N_2 is bubbled through the solution for 2 minutes to eliminate O_2 from the added liquid. The flow of N_2 is stopped and allowed to pass over the surface of the solution. Now the current is noted. This procedure is repeated until the whole of the solution is titrated. The diffusion current is plotted against the volume of reagent, in which the intersection of these two lines gives the end point.

Titrations with Rotating Pt Electrode

The dropping mercury electrode cannot be used at more positive potentials (namely above +0.4 V vs SCE) because of the oxidation of the Hg. Hence it can be replaced by an inert platinum electrode (can be used up to 1.1 V). The attainment of diffusion current is slow with a stationary Pt electrode and it can be overcome by rotating the Pt electrode at constant speed. The larger currents (about 20 times those at the DME) attained with the rotating platinum electrode allow correspondingly smaller currents to be measured without loss of accuracy and thus very dilute solutions (10^{-4}M) may be titrated. In order to get a linear relation between the current and the volume of reagent, the speed of the electrode must be kept constant (600 rpm (revolution per minute) is suitable).

The electrode is constructed from a standard mercury seal and is shown in Figure 4.5. About 5 mm of platinum wire (0.5 mm diameter) protrudes from the wall of a glass tube of length 6 mm; the latter is bent at an angle, approaching a right angle, a short distance from the lower end. Electrical connection is made to the electrode by an amalgamated Cu wire passing through the tubing to the mercury covering the sealed-in platinum wire; the upper end of the copper wire passes through a small hole blown in the stem of the stirrer and dips into mercury contained in the "mercury seal". A wire from the latter is connected to the source of applied voltage. The tubing forms the stem of the electrode, which is rotated at a constant speed of 600 rpm.

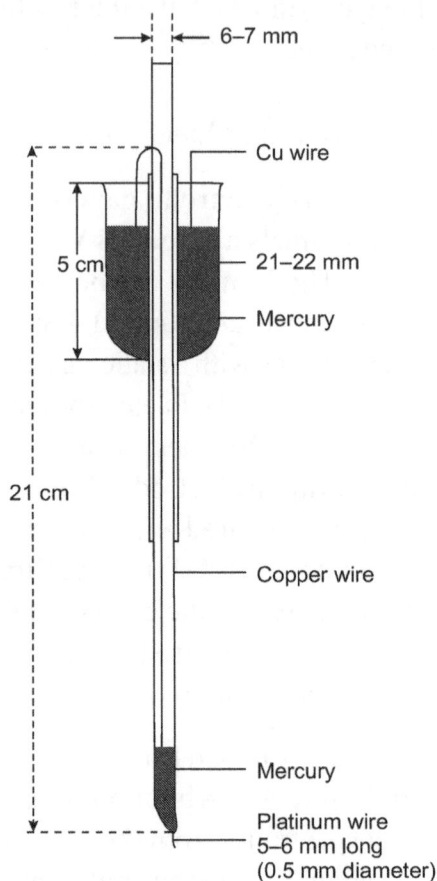

Figure 4.6   Rotating platinum microelectrode

BIAMPEROMETRIC TITRATIONS

Titrations are performed in a uniformly stirred solution by using two small but similar platinum electrodes to which a small emf (10–100 mV) is applied. The end point is shown by either the disappearance or the appearance of current flowing between two electrodes. The only requirement for this experiment is that a reversible oxidation–reduction system be present either before or after the end point.

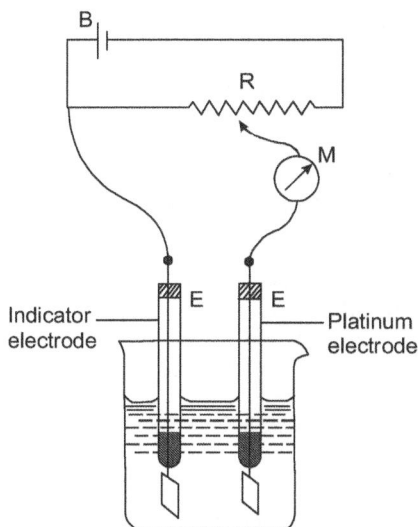

Figure 4.7   A simple cell set-up for biamperometry

Figure 4.7 shows a simple set-up for the biamperometric technique. A 3-volt battery (B), microammeter (M), 500-ohm resistor (R), and platinum electrodes (E, E) are connected in a circuit. The potentiometer is set so that there is a potential drop of about 100 mV across the electrodes. In a titration with two indicator electrodes and in which the reactant involves a reversible system ($I_2 + 2e^- \rightleftharpoons 2I$), an appreciable current flows through the cell. The amount of oxidized form reduced at the cathode is equal to that of the reduced form oxidized at the anode. Both the electrodes are depolarized until the oxidized component or the reduced component of the system has been consumed by a titrant. After the end point, only one electrode remains depolarized if the titrant does not involve a reversible system. Current thus flows until the end point—at or after the end point, the current is zero. The complementary type of end point is obtained in the titration of an irreversible couple by a reversible couple. The current is low before the end point and rapidly increases after the end point.

Some examples of amperometric titrations are given in Table 4.1.

Table 4.1   Examples of amperometric titrations

Titrant	Electrode	Species determined
Complexation reaction		
EDTA	DME	Almost all metallic ions
Precipitation reaction		
DMG	DME	Ni^{2+}
Lead nitrate	DME	SO_4^{2-}, MoO_4^{2-}, F^-
$Hg(NO_3)_2$	DME	I^-
$AgNO_3$	Rot. Pt	Cl^-, Br^-, I^-, CN^-, thiols
Sodium tetraphenylborate	Graphite	K^+
$Th(NO_3)_4$	DME	F^-
$K_2Cr_2O_7$	DME	Pb^{2+}, Ba^{2+}
Redox reaction		
I_2	Rot.Pt	As (III), $Na_2S_2O_3$
$KBr/KBrO_3$	Rot.Pt	As (III), Sb(III), N_2H_4
Additions	Rot.Pt	Alkenes
Substitutions	Rot.Pt	Phenols, aromatic amines

Advantages of Amperometric Titrations

1. Titrations can be carried out rapidly since the end point is found graphically.

2. Titrations can be carried out in the cases where the solubility relations (such that potentiometric or visual indicator methods) are unsatisfactory.

3. Very dilute solutions (10^{-4}M) can be titrated whereas potentiometric titrations no longer yield accurate results.

4. "Foreign" salts may frequently be present without interference and are usually added as the supporting electrolyte in order to eliminate the migration current.

APPLICATIONS

Qualitative determination   As depicted in Figure 4.2, with a thin-layer cell with two working electrodes of parallel mode, one can monitor the current at two different potentials. The ratio of output of these two electrodes indicate peak purity. From these, a general idea of the group undergoing the electrochemical oxidation (or reduction) reaction is obtained.

Quantitative determination

❖ The concentrations of electroinactive compounds (within practical potential limit) like acetylcholine and amino acids are determined by derivatizing to electroactive compound.

❖ Estimation of SO_4^{2-} is not possible by potentiometric methods due to lack of suitable indicator and by conductometric method in the presence of foreign electrolyte. But the estimation of some is possible by amperometric method.

❖ Fluoride ion can be estimated with thorium and lanthanum nitrate.

Problems

1. Based on the curves in the figure below, consider the titration of Sn^{2+} with I_2 using one-electrode amperometry. Sketch the resulting titration curve for a platinum indicator electrode maintained at (a) +0.2 V, (b) –0.1 V, and (c) –0.4 V vs SCE

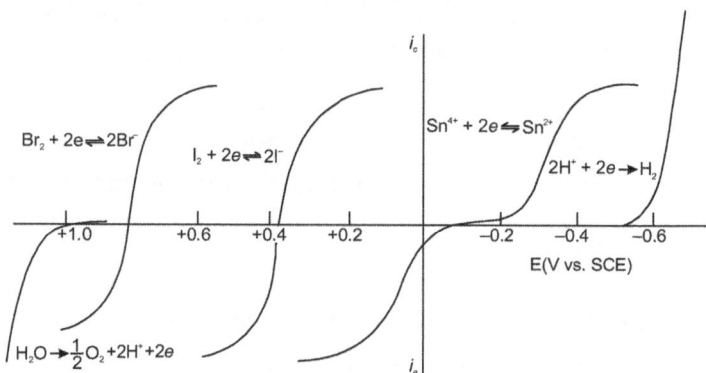

2. Based on the curves in the above figure how could one determine a mixture of Br_2 and I_2 by titration with Sn^{2+} using one-electrode amperometry? Sketch the current-potential curve that would be obtained for various stages of titration and the amperometry titration curves that would result from the method you propose. Sketch the titration curves of a mixture of Br_2 and I_2 by titration with Sn^{2+} using two-electrode amperometry with an impressed voltage of 100 mV.

3. Iodine is to be titrated coulometrically at constant current at silver electrode. The sample is 1.0 mM sodium iodide contained in a pH 4 acetic acid solutions with 0.1 M sodium acetate with a total volume of 50 ml. Describe the course of the titration. What generating current do you recommend and what total titration time is expected?

4. Iodide can be determined by controlled potential oxidation to iodine at a platinum electrode. What potential should be used for this oxidation? How many coulombs will be passed?

5. Consider carrying out a one-electron amperometric titration for the system Fe^{2+}- Ce^{2+} as shown in the following figure at several widely different potentials and sketch the amperometric titration curves that result. Consider in each case, situations in which (a) the mass transfer coefficients for all species are equal and (b) those for iron ions are 25% larger than those for certain species. Which curves would be useful in a practical titration?

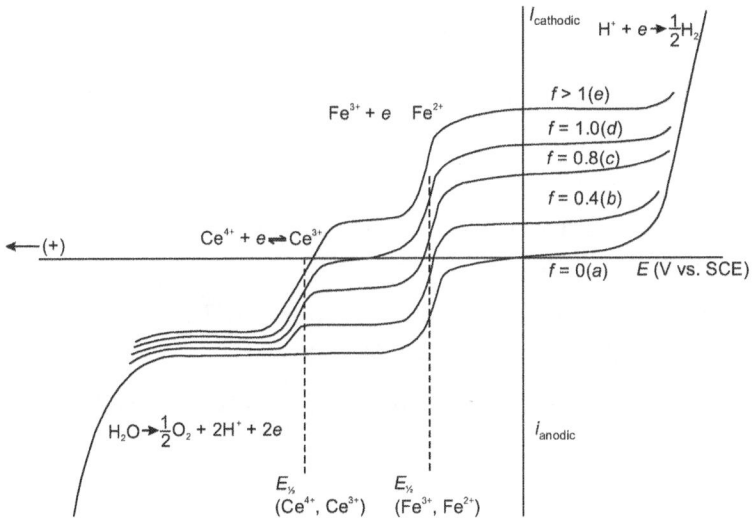

References

Fleet, B. and Gunasingham, H. (1992) "Electrochemical Sensors for Monitoring Environmental Pollutants." *Talanta*. 39, 1449.

Frew, J.E. and Green, M.J. (1988). "Biosensors for clinical analysis." *Analytical Proceedings*. 25, 276.

Frew, J.E. and Green, M.J. (1989). "Amperometric biosensors." *Analytical Proceedings*. 26, 334.

Hart, J.P. and Wring, S.A. (1994). "Screen-printed voltammetric and amperometric electrochemical sensors for decentralised testing." *Electroanalysis*. 6, 617.

Kauffmann, J.M. and Guilbault, G.G. (1992). "Enzyme Electrode Biosensors: Theory and Applications." In: *Bioanalytical Applications of Enzymes*. Suelter, C.H. (ed.). Vol.36. Wiley, New York. p.63.

Murray, R.W., Ewing, A.G. and Durst, R.A. (1987). "Chemically Modified Electrodes: Molecular Design for Electroanalysis." *Analytical Chemistry*. 59, 379A.

Turner, A.P.F. (ed.). *Biosensors and Bioelectronics*. Elsevier Science Publishing Co. Inc., New York.

Wang, J. (ed.). *Electroanalysis*. VCH Publishers Inc., New York.

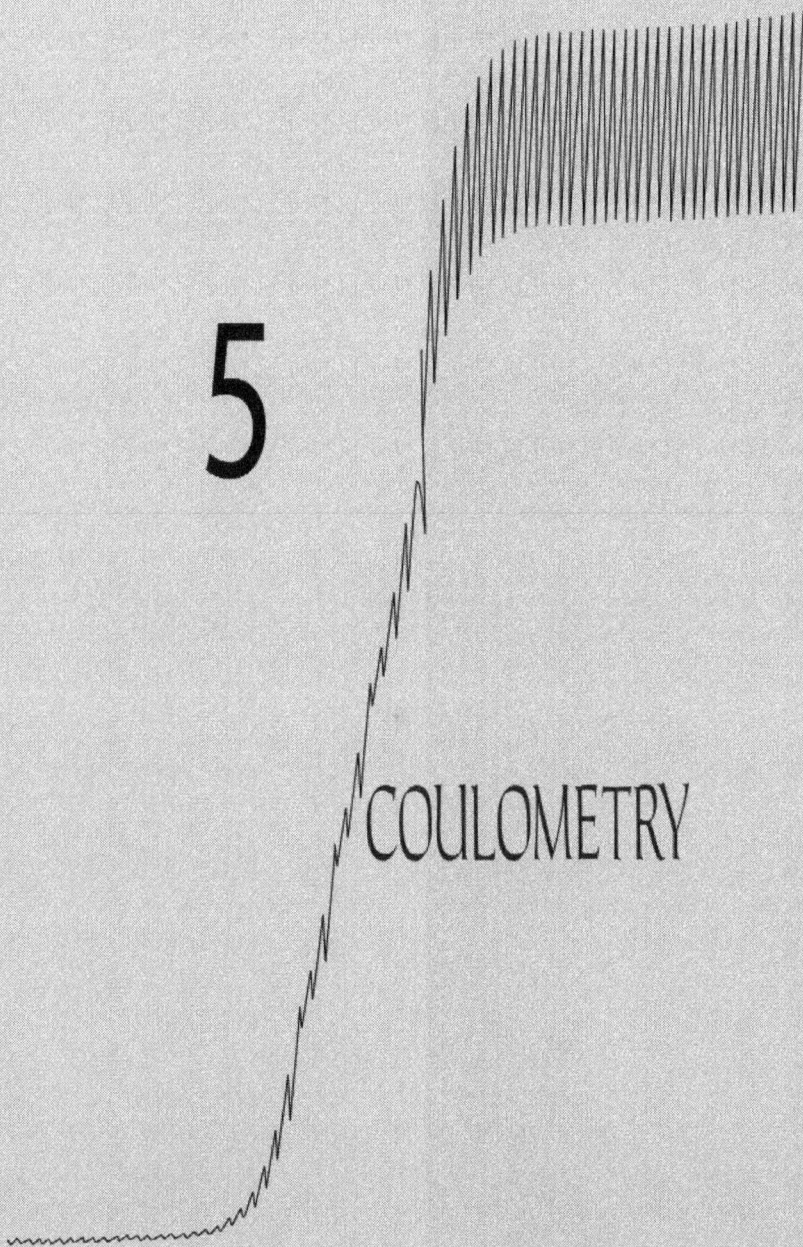

5

COULOMETRY

Coulometry is an electroanalytical technique that determines the amount of matter transformed during an electrolysis reaction by measuring the amount of electricity (in coulombs) consumed or produced. This technique is applicable to redox reactions in which electrons are transferred from one molecule to another. The reaction is controlled by applying an electrical potential, and the amount of electricity (i.e., no. of electrons) needed to complete the reaction is the main measurement.

This technique is as accurate as other analytical techniques that are used to carry out similar analyses. But this coulometric technique is a laborious process. Since coulometric methods do not measure directly the amount of matter transformed in the redox reaction, it is important that the reaction maintains a stoichiometrical relationship. Thus, each electron transferred must correspond to a known number of molecules in the reaction. The amount of current can be determined by using the following equation.

$$Q = It$$

where,

Q is the amount of electricity in coulombs,

I is the current in amperes and

t is the time in seconds.

Q is then related to the reaction by the stoichiometric ratio. As the reaction progresses, the concentration of the molecules being converted will decrease. The electrical potential must be increased to maintain the direction of the reaction as the concentration in the sample may change.

The methods can be classified as controlled current methods and controlled potential methods. Thus in the controlled potential technique, the potential of the working electrode is maintained constant with respect to a reference electrode. Since the potential of the working electrode is a basic variable that controls the degree of completion of an electrolytic process in most cases, controlled potential techniques are usually the most desirable for bulk electrolysis. However, these methods require potentiostats with large output current and voltage capabilities as well as stable reference electrodes, carefully placed to minimize uncompensated resistance effects. Auxiliary electrode placed to provide a fairly uniform current distribution across the surface of the working electrode is usually desirable, and it is often placed in a separate compartment isolated from the working-electrode compartment by a sintered-glass disc, ion-exchange membrane or other separator. A related technique in which bulk electrolysis occurs without the use of an external power supply is implemented by choosing a counter electrode that makes the entire cell a galvanaic cell, so that some degree of potential control occurs during discharge. This technique called internal electrolysis is now rarely used.

In controlled current techniques, the current passing through the cell is held constant (or sometimes programmed to change with time). Although the techniques frequently involve simple instrumentation than controlled potential methods, it requires either a special set of chemical conditions in the cell or some specific detection methods to signal the completion of the electrolysis and to ensure 100% current efficiency.

CONTROLLED POTENTIAL METHODS

Consider the electrolysis of an oxidant O initially present in a bulk solution at a concentration C_0, by the reaction $O + ne^- \rightleftharpoons R$ at an electrode of area A held at a potential, E_c corresponding to the limiting current region.

The current at any time is given by

$$i(t) = nFAm_0 \cdot c_0(t)$$

The current also indicates the total rate of consumption of O, $\dfrac{dN_o}{dt}$ due to electrolysis

$$i(t) = -nF\left(\frac{dN_o}{dt}\right)$$

where, $\dfrac{dN_o}{dt}$ is the change in total number of moles of O in the system with respect to time.

By relating the above two equations, we obtain

$$\frac{dC_o(t)}{dt} = -\left(\frac{m_o A}{V}\right)C_o(t)$$

with the initial condition $C_o(t) = C_o(0)$ at $t = 0$.

Thus

$$C_o(t) = C_o(0)\exp\left(\frac{-m_o A}{V}\right)t$$

or

$$i(t) = i(0)\exp\left(\frac{-m_o A}{V}\right)t$$

where $i(0)$ is the initial current. Thus a controlled potential bulk electrolysis is like a first-order reaction, with the concentration and the current decaying exponentially with time during the electrolysis and attaining the background (residual) current.

The total quantity of electricity $Q(t)$ (in coulombs) consumed in the electrolysis is given by the area under the i–t curve

$$Q(t) = \int_0^t i(t)dt$$

Electrolysis at controlled potential is the most efficient method of carrying out a bulk electrolysis, because the current is always maintained at the maximum value consistent with 100% current efficiency.

Coulometric Measurements

In the controlled potential coulometry, the total number of coulombs consumed in an electrolysis is used to determine the amount of the substance electrolysed. To enable a coulometric method, the electrode reaction must satisfy the following requirements.

i. it must be of known stoichiometry.

ii. it must be a single reaction or at least have no side reactions of different stoichiometry.

iii. it must occur with close to 100% current efficiency. A block diagram of the apparatus used in controlled potential coulometry is shown in Figure 5.1.

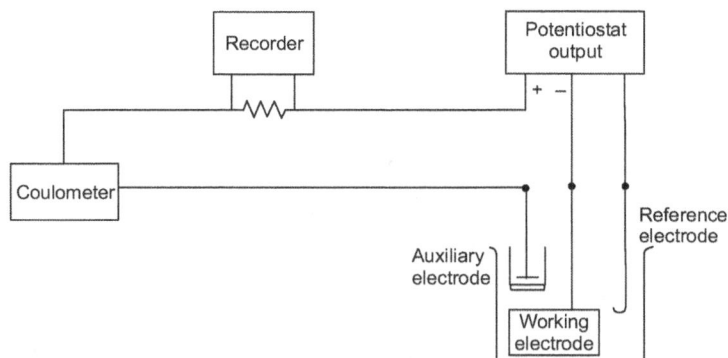

Figure 5.1   Apparatus used in controlled potential coulometry

The potentiostats used generally have an output power of 100 W. Modern electronic potentiostats employing operational amplifier control circuits at solid-state output devices are available; these are more convenient to use and show much faster response time. The current is monitored during the electrolysis with a strip chart recorder, so that the background current can be determined and the completion of electrolysis can be observed. The shape of the *i–t* curve can be diagnostic of the mechanisms of the electrode reactions and instrumental problems.

There have been many applications of controlled potential coulometry to analysis. It is a useful method for studying the mechanisms of electrode reactions and for determining the *n* values for an electrode reaction without prior knowledge of electrode area or diffusion current.

CONTROLLED CURRENT METHODS

A constant coulometric method is attractive because a stable constant current source is easy to construct and the total number of coulombs consumed in an electrolysis can be calculated from the duration of the electrolysis τ which is given by

$$Q = i_{app}\tau$$

However to use a coulometric method for determination, the reaction of interest must proceed with close to 100% efficiency. This can be achieved and easily understood by the following method.

Coulometric Measurements

Consider the coulometric determination of Fe^{2+} by oxidation at a Pt electrode to Fe^{3+} in an H_2SO_4 medium. If a constant current is applied to the Pt anode, i_l for Fe^{2+} oxidation falls below i_{app}, the current efficiency would fall below 100% and part of the applied current would go to evolution of O_2 (side reaction). If Ce^{3+} is added to the solution, when the current efficiency for the direct oxidation of Fe^{2+} falls below 100%, the next process to occur is

$$Ce^{3+} \longrightarrow Ce^{4+} + e^-$$

The Ce^{4+} produced is capable of oxidizing any Fe^{2+} remaining in the bulk solution by the fast reaction.

$$Ce^{4+} + Fe^{2+} \longrightarrow Fe^{3+} + Ce^{3+}$$

Thus Fe^{2+} is indirectly oxidized to Fe^{3+}, and the titration efficiency for the oxidation of Fe^{2+} is maintained. This technique is called coulometric titration. Neither the current nor the potential of the working electrode is a good indicator of the course of the reaction, hence an end point detection technique must be used to indicate the completion of the reaction.

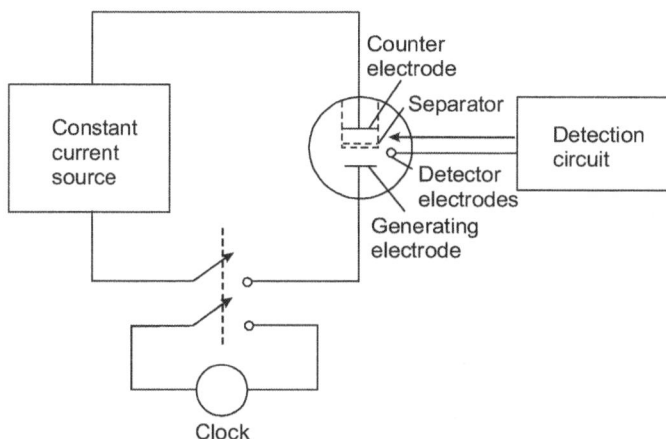

Figure 5.2 Apparatus used in coulometric titration

A block diagram of the apparatus used in coulometric titration is shown in Figure 5.2. The cell is composed of a working electrode and an auxiliary electrode in separate compartments. Indicator electrodes suited to the particular end point detection technique are also located within the cell. The constant current source can be a high-voltage (~400 V) power supply and a bank of resistors. This will produce a constant current as long as the reversible cell potential and cell resistance are small compared to the applied voltage and circuit resistance. Whenever current is switched to the cell, a timer is actuated, so that the total electrolysis time can be recorded.

The current density range for generation of the titrant can be determined by taking *i–e* curves of the supporting electrolyte system with and without the titrant. A plot of current efficiency as a function of current density can be prepared and from this the optimum region for titrant generation can be established. The current to be used is selected by taking into consideration the amount of substance to be determined and a convenient electrolysis time. Then an electrode area is calculated to give the needed current density.

APPLICATIONS

i. This technique has been used to determine over 50 elements. It is used very frequently for the determination of uranium and plutonium.

ii. It determines the amount of O_2 in a sample by using a cadmium electrode and a porous Ag electrode.

iii. Ions such as halides, Zn^{2+} and mercaptans can be determined by complex formation or precipitation in the presence of anodically generated silver ions.

iv. The oxidation of cerium in sulphuric acid medium can be used to characterize various systems such as Ti, Fe, and uranium.

v. The amount of water in a sample can be determined in the Karl Fischer reaction. This reaction uses coulometric titration to determine the amount of water in myriad substances such as butter, sugar, cheese, paper and petroleum.

vi. The thickness of metal film coated on copper wire or steel can be determined. This is performed by measuring the quantity of electricity needed to dissolve a well-defined area of the coating. The film thickness is proportional to the constant current i, the molecular weight of the metal, the density of the metal and the surface area.

vii. The coulometric determination is used in blood glucose monitoring devices, and controlled potential coulometric method is used to determine certain carcinogens and drugs.

Problems

1. Fifty ml of $ZnSO_4$ solution is transferred to an electrolytic cell with a mercury cathode and enough solid potassium nitrate is added to make the solution 0.1 M in KNO_3. The electrolysis of Zn^{2+} is carried to completion to a potential of −1.3 V vs SCE with the passage of 241 coulombs of electricity. Calculate the initial concentration of zinc ion.

2. When a solution of volume 100 cm² containing metal ion, at a concentration 0.01 M is electrolysed with a rapid scan at a large-area rotating disc electrode, a limiting current of 193 mA is observed for reduction to metal M. Calculate the value of $m_{M^{2+}}$ the mass transport coefficient, in cm/sec. If electrolysis of the solution is carried out at this electrode at controlled potential in the limiting current region, what time will be required for 99% of the M^{2+} to be placed out? How many coulombs will be required for this electrolysis?

3. If the solution in previous problem is electrolysed at a constant current of 80 mA under the same conditions: (a) What is the concentration of M^{2+} remaining in solution when the current efficiency drops below 100%? (b) How long does it take to reach this point? (c) How many coulombs have been passed to this point? How much longer will it take to decrease the M^{2+} concentration to 1.1% of its initial value? What is the overall current efficiency for removal of 99.9% of M^{2+} by this constant current electrolysis?

References

DeFord, D. and Miller, J.W. (1963). In: *Treatise on Analytical Chemistry*. Part I, Vol. 4. Kolthoff, I. M. and Elving, P.J. (eds.). Wiley-Interscience, New York. Ch. 49.

Harrar, J.E. (1975). *Electroanal. Chem.* 8, 1.

Kies, H.L. (1962). *J. Electroanal. Chem.* 4, 257.

Lingane, J.J. (1958). *Electroanalytical Chemistry*, 2nd edn., Wiley-Interscience, New York. Ch. 13–21.

Rechnitz, G.A. (1963). *Controlled Potential Analysis*. Pergamon, New York.

GLOSSARY

AC   Alternating current.

Activity   Concentration expressed in terms of molality.

Activity coefficient   An interionic attraction between ions.

Amperometry   Technique which measures variation in current measured as a function of time or volume of titrant, when constant potential is applied between the working and counter electrodes.

Anodic peak current   Current corresponding to anodic peak.

Anodic peak potential   Potential corresponding to anodic peak.

Anodic shift   Potential moving towards anodic direction.

Biocatalytic electrode   An electrode or membrane showing reversibility with biologically important molecules.

Cathodic peak current   Current corresponding to cathodic peak.

Cathodic peak potential   Potential corresponding to cathodic peak.

Cathodic shift   Potential moving towards cathodic direction.

Cell constant   Ratio of specific conductivity to observed conductivity.

Chronopotentiometry   Application of constant current between working and auxiliary electrodes and monitoring variation in potential with time.

Conductance   Reciprocal of resistance.

Coulometry   Amount of electricity or coulombs produced or consumed in redox reaction when potential is applied.

Current ratio   Ratio of anodic peak current to cathodic peak current.

Current function   Ratio of peak current to square root of scan rate.

DC   Direct current.

Decomposition potential   Potential at which the electroactive species undergoes oxidation or reduction.

Diffusion current   Current carried by ion when it diffuses from region of higher concentration to lower concentration.

Dissociation constant   The fraction of current carried by an electrolyte.

DME   Dropping mercury electrode.

Drop time   Time taken by a drop to fall.

Electroactive species   A substance or species undergoing redox reaction or electron transfer reaction.

Electrolyte   An organic or inorganic salt dissociating into ions in solvent.

EMF (electromotive force)   The flow of current between the two electrodes of different potentials.

Equitransferent electrolyte   A salt containing cations and anions of equal mobility.

Equivalent conductance   Conductance of one equivalent of an electrolyte.

Faradaic current   Current carried by an electron transfer across the electrode–electrolyte interface and which obeys Faraday's law.

Faraday's first law   This law states that the number of ions discharged at an electrode is directly proportional to the total quantity of electricity passed through the solution.

Faraday's second law   This law states that the number of faradays of electric charge required to discharge one mole of substance at an electrode is equal to the number of "excess" elementary charges on that ion.

Fick's first law   Fick's first law is used in steady-state diffusion, i.e., when the concentration within the diffusion volume does not change with respect to time.

Fick's second law   Fick's second law is used in non-steady or continually changing state diffusion, i.e., when the concentration within the diffusion volume changes with respect to time.

Galvanaic cell An electrochemical cell that produces electrical energy at the expense of chemical energy.

Gas sensing electrode An electrode that measures the concentration of gases or ions whose conjugate acid or base is a gaseous species.

Glass electrode A special type of glass having low melting point and good electrical conductivity and shows reversibility with respect to hydrogen ion concentration.

Half wave potential Potential at half of diffusion current value.

Ion selective electrode An electrode selective to one particular ion.

Ionophore A liquid ion exchange material.

Kinetic current Current is purely determined by the rate of chemical reaction.

Liquid junction potential Potential difference that exists when two different solutions containing different ions come in contact with each other.

Mobility The ease with which an ion can move.

Molar conductance Conductance of one mole of an electrolyte.

Non-faradaic current No electron transferred across the electrode and electrolyte interface and current is carried by means of adsorption or desorption process which are non-faradaic process.

Ohm's law This law states that the strength of current flowing through a conductor is directly proportional to the potential difference across the electrode and inversely proportional to the resistance of the conductor.

Over potential Potential applied over and above the equilibrium potential.

Polarography Technique that measures the diffusion current when potential is applied between a dropping mercury electrode and a counter electrode.

Reference electrode An electrode whose standard reduction potential is constant with respect to temperature and the potential of any electrode is measured in comparison with the reference electrode.

Resistance It is directly proportional to the length and inversely proportional to the area of the electrode.

Scan rate Potential (mV or V) applied per second.

Specific conductance Conductance of 1 ml of an electrolyte.

Standard oxidation potential The tendency of a metal to lose electron when it is dipped in a solution of its own salt solution of unit activity at 25°C.

Standard reduction potential The tendency of a metal to gain electron when it is dipped in a solution of its own salt solution of unit activity at 25°C.

Transport number Fraction of total current carried by an electrolyte.

Voltaic cell An electrochemical cell that produces chemical energy at the expense of electrical energy.

INDEX